Wildflowers of the Carolina Lowcountry and Lower Pee Dee

Wildflowers of the Carolina Lowcountry and Lower Pee Dee

Richard D. Porcher

University of South Carolina Press

All photographs are by the author.

Published in Columbia, South Carolina by the
University of South Carolina Press

Printed in Canada

99 98 97 96 95   5 4 3 2 1

Library of Congress Cataloging-in-Publication Data

Porcher, Richard D. (Richard Dwight)
Wildflowers of the Carolina lowcountry and lower Pee Dee / Richard D. Porcher.
p. cm.
Includes bibliographical references and index.
ISBN 1-57003-094-4 (cl) ISBN 1-57003-027-8 (pbk)
1. Wild flowers—South Carolina—Identification. 2. Wild flowers—South Carolina—Pictorial works. 3. Wild flowers—South Carolina—Geographical distribution. I. Title.
QK185.P83 1995
582.13' 09757—dc20   94-18771

TO REX K. SKANCHY,

a member of The Citadel Family for thirteen years. He was my friend and the Biology Department's friend. Most important, he was a friend to any student at The Citadel who needed one. The Rex K. Skanchy Biology Scholarship was established in 1993 by the Biology Department and his friends in Rex's memory. All author's royalties from this book will go to the scholarship.

Brief Contents

Contents

Figures

Acknowledgments

I am grateful to numerous people and organizations who helped in the preparation and publication of this book and who also made it such a pleasant and enjoyable challenge. I am aware that a book of this type, although new in many ways, is never the sole product of the author. Rather it was built on the knowledge and data accumulated throughout the history of the discipline by many dedicated and talented botanists. Just as this book has made use of this knowledge, I hope this material has added to the field of botany and will be used in future books on the Lowcountry and South Carolina.

No book of this magnitude has a more important contribution than the critical review for technical, scientific accuracy. I am deeply grateful to two of my professional colleagues, Dr. Douglas A. Rayner of Wofford College and Dr. Joseph N. Pinson of Coastal Carolina University, for their review of the manuscript and helpful comments and appropriate questions throughout the project.

I gratefully acknowledge the valuable editorial assistance and suggestions of Angela Williams, director of The Citadel Writing Center, who spent many hours reviewing and editing the manuscript.

Again I had the pleasure of working with Jane Lareau, who helped write and organize parts of the introduction. We had collaborated on the text of *Lowcountry: The Natural Landscape* and for this volume there was no question from whom I would get help with polishing the rough edges.

It is especially appropriate to acknowledge David V. Hamilton and the Tradd Street Press of Charleston, South Carolina. David initially approached me with the challenge of publishing a book on Lowcountry wildflowers, which I immediately accepted. Even though Tradd Street Press was unable to publish the book, his early financial support and encouragement made the project successful.

I owe a special thanks to my friend Roy Rooks for his companionship and technical assistance photographing wildflowers on numerous trips in the field. Even though this was the fun part of the project, field photography in the Lowcountry summers sometimes requires an extra effort, one that Roy provided uncomplainingly.

I am indebted to the following for financial support: The Citadel Development Foundation, the Charleston Garden Club, the Biology Department of The Citadel, the S.C. Coastal Conservation League, William Cain, Jr., M.D., the Bio-Cid Club and the Tau Nu Chapter of Tri-Beta of The Citadel Biology Department, and the S.C. National Bank of Charleston as trustee of the John D. Muller Charitable Trust.

It has been a pleasure to work with the staff of the University of South Carolina Press. I would like to thank Rebecca Blakeney for her design expertise and Jamie Browne for her editorial contributions. I especially thank Warren Slesinger, acquisitions editor, for his helpful contributions and guidance.

I also thank Diane Raeke, a medical illustrator from Charleston, S.C., for the illustrations of plant structures and Judy Burress for the map of the Lowcountry and Lower Pee Dee.

Gay Johnson became my friend and wife during the final work on the book; I thank her for her patience and companionship during the last, hectic year.

And, finally, I thank my mentor who started me on my botanical journey through the Carolina Lowcountry, my major professor, Dr. Wade T. Batson, Professor Emeritus at the University of South Carolina. My first day in the field with him as a graduate student was the acorn that matured into the field botanist that accepted the challenge to produce this book.

Wildflowers of the Carolina Lowcountry and Lower Pee Dee

MAP OF THE LOWCOUNTRY AND LOWER PEE DEE

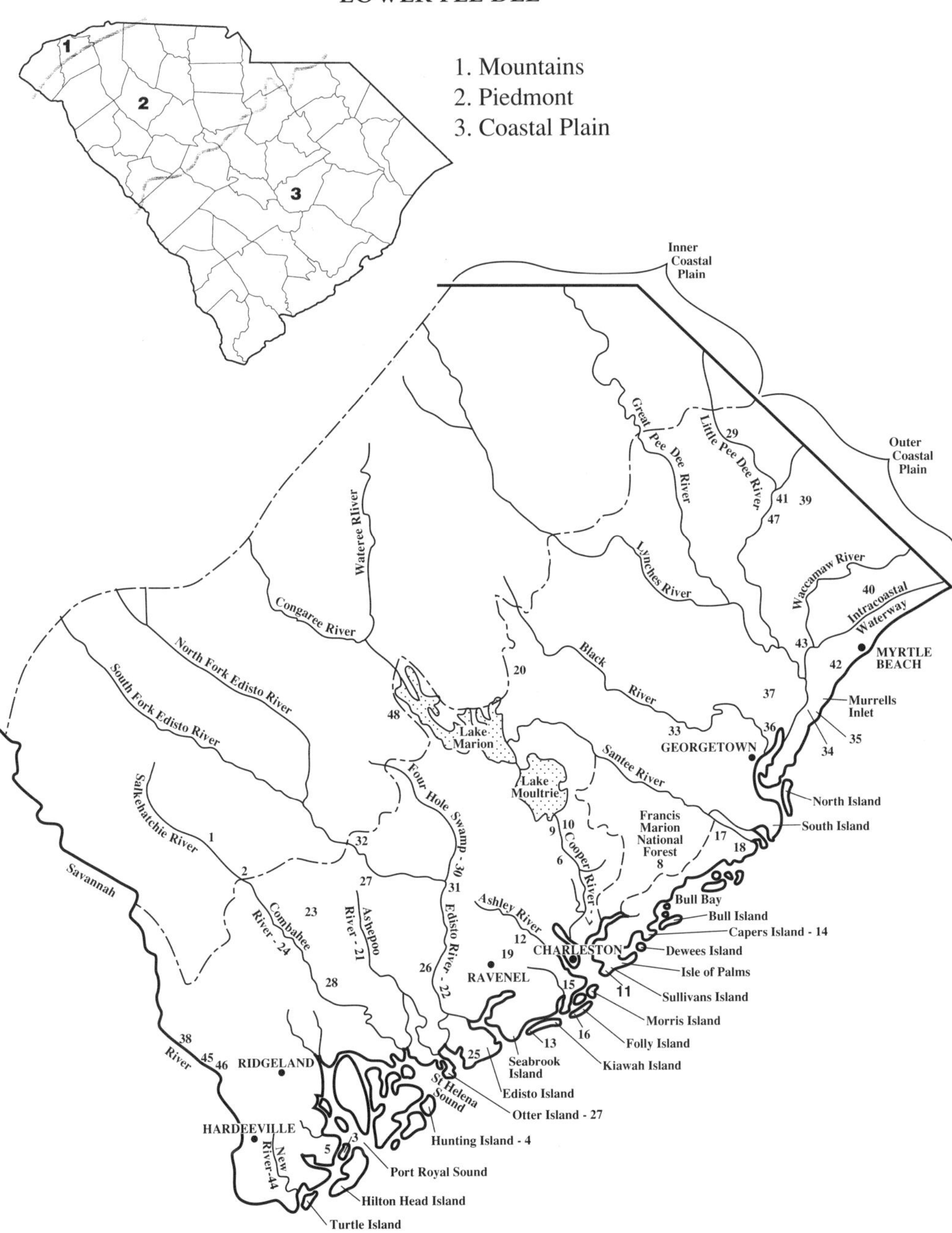

TABLE OF WILDFLOWER SITES AND ASSOCIATED GARDENS

Site	Coastal Beaches	Coastal Dunes	Maritime Shrub Thickets	Maritime Forests	Salt Shrub Thickets	Salt Marshes	Salt Flats	Shell Mounds	Brackish Marshes	Tidal Freshwater Marshes	Tidal Freshwater Swamps	Stream Banks	Freshwater Aquatics	Pocosins	Bald Cypress–Tupelo Gum Sw. For.	Hardwood Bottoms	Marl Forests	Beech Forests	Oak-Hickory Forests	Sandy, Dry, Open Woods	Xeric Sandhills	Longleaf Pine Flatwoods	Longleaf Pine Savannas	Pine–Saw Palmetto Flatwoods	Pine–Mixed Hardwood Forests	Pond Cypress Savannas	Pond Cypress–Swamp Gum Sw. For.	Ruderal Gardens
Accreted Beach at Sullivans Island (11)	■	■	■																									
Ashepoo River (21)						■			■	■	■																	
Audubon Swamp Garden at Magnolia Gardens (12)													■															■
Beachwalker County Park (13)	■	■	■		■	■	■																					
Bennett's Bay Heritage Preserve (20)														■							■							
Black River Swamp Heritage Preserve (33)										■	■																	
Bluff Plantation Wildlife Sanctuary (6)										■	■		■		■				■			■	■		■			■
Brookgreen Gardens Xeric Sandhills (34)																					■							
Capers Island Heritage Preserve (14)	■	■	■	■	■	■	■																					■
Cartwheel Bay Heritage Preserve (39)														■							■	■	■					
Cathedral Bay Heritage Preserve (1)																										■		
Colleton State Park (22)															■	■												■
Combahee River (24)						■			■	■	■																	
Cooper River (7)						■			■	■	■		■															
Crosby Oxypolis Heritage Preserve (23)														■													■	
Dill Wildlife Refuge (15)						■	■																		■			■
Edisto Beach State Park (25)	■	■	■	■	■	■	■	■																				■
Edisto Nature Trail (26)										■	■					■		■							■			■
Folly Beach County Park (16)	■	■	■		■	■	■																					■
Francis Beidler Forest in Four Hole's Swamp (30)															■	■												
Francis Marion National Forest (8)				■	■	■	■	■	■	■	■	■	■	■	■	■		■		■	■	■	■		■	■	■	■
Givhans Ferry State Park (31)															■	■	■	■										■
Green's Shell Enclosure Heritage Preserve (3)				■	■	■	■	■																				■
Hampton Plantation State Park (17)													■	■								■	■		■		■	■

Site	Coastal Beaches	Coastal Dunes	Maritime Shrub Thickets	Maritime Forests	Salt Shrub Thickets	Salt Marshes	Salt Flats	Shell Mounds	Brackish Marshes	Tidal Freshwater Marshes	Tidal Freshwater Swamps	Stream Banks	Freshwater Aquatics	Pocosins	Bald Cypress–Tupelo Gum Sw. For.	Hardwood Bottoms	Marl Forests	Beech Forests	Oak-Hickory Forests	Sandy, Dry, Open Woods	Xeric Sandhills	Longleaf Pine Flatwoods	Longleaf Pine Savannas	Pine–Saw Palmetto Flatwoods	Pine–Mixed Hardwood Forests	Pond Cypress Savannas	Pond Cypress–Swamp Gum Sw. For.	Ruderal Gardens
Hunting Island State Park (4)	■	■	■	■	■	■	■																					■
Huntington Beach State Park (35)	■	■	■	■	■	■	■						■															■
Lewis Ocean Bay Heritage Preserve (40)														■							■	■	■					
Little Pee Dee River Heritage Preserves (47)												■			■	■												
Little Pee Dee River Her. Pres. (Vaughn Tract) (41)														■							■							
Little Pee Dee State Park and Bay (29)													■	■							■				■			■
McAlhany Preserve (32)													■		■	■		■	■									■
Myrtle Beach State Park (42)	■	■	■	■	■	■	■																					■
New River (44)						■			■	■	■																	
Old Santee Canal State Park (9)										■	■		■				■	■	■									■
Otter Island (St. Helena Sound Her. Pres.) (27)	■	■	■	■	■	■	■	■																				
Rivers Bridge State Park (2)															■													■
Samworth W.M.A. and Great Pee Dee River (36)										■	■																	
Santee Coastal Reserve (18)				■					■				■		■							■	■		■	■	■	■
Santee State Park (48)																	■	■	■									■
S.C. Highway 119 (45)																				■	■							■
Tea Farm County Park (19)					■	■									■			■	■						■			■
Tillman Sand Ridge Heritage Preserve (46)															■	■				■	■							
U.S. Highway 701 (37)														■								■	■					■
Victoria Bluff Heritage Preserve (5)				■									■	■										■				
Waccamaw River Preserve (43)										■	■																	
Wadboo Creek Marl Forest (10)										■	■						■											
Webb Wildlife Center (38)													■		■	■			■			■	■		■			■
Westvaco's Bluff Trail (28)													■		■										■			■

PART ONE

Wildflower Gardens of the Lowcountry and Lower Pee Dee

Introduction

PURPOSE AND SCOPE

Welcome to the wildflower gardens of South Carolina's Lowcountry and Lower Pee Dee regions. This volume includes 437 species of wildflowers and serves the needs of both the amateur and professional botanists. Because many of the wildflowers and natural gardens occur in the coastal area of North Carolina and Georgia, and further inland toward the piedmont of South Carolina, this book also serves as a supplementary guide for these areas.

I am pleased to be able to offer you this book which is designed to go beyond simply identifying the species you are viewing. As the author, drawing on several decades of studying the natural history and ecology of this region, I share much of what I have learned about South Carolina's richest and most beautiful wildflower habitats.

First, this book helps you identify wildflowers in a very different way. Flowers are grouped for identification not by their color or family group, as is commonly done, but by where you will most likely find them growing—in their natural gardens.

For example, you are near the ocean and want to identify a small, daisylike flower. You will not look at the daisies, nor will you have to thumb through all the yellow flowers. Instead, you will go to "Seaside Gardens" and look under the section labeled "Coastal Beaches" or "Coastal Dunes," where your specimen is obviously thriving. Likewise, if you are in a tidal swamp, you would turn to "River Gardens" and check under "Tidal Freshwater Swamp Forests"; or if you are in a forest of towering pines, you would flip to "Pineland Gardens" and search through "Longleaf Pine Flatwoods" or "Longleaf Pine Savannas."

Included in the wildflower gardens are the ruderal gardens such as roadsides, yards, sidewalks, and disturbed woods where the majority of weedy native and naturalized wildflowers occur. This gives those less inclined to venture far from home a ready guide to their backyards. Photographs of seventy-eight species in this garden are included.

This wildflower book is also different because it provides a guide to special locations in the Lowcountry and Lower Pee Dee where you can find representative samplings of the twenty-seven natural gardens described. I have chosen forty-eight of the best sites where these gardens can be found: from the seaside gardens at Hunting Island State Park, to the blackwaters of the Ashepoo River and the marsh flowers on the banks, to the woods of the Francis Marion National Forest with the native wildflowers of the longleaf pine savannas. All of these sites are open to the public, and clear directions make them easy to find. In addition, contacts are provided to allow you to obtain more information on the sites.

This book also offers a special section on the ecology of the natural gardens. Wildflowers do not grow in a vacuum. In fact, nothing is more important to their existence than the habitat you will find them in. Several essays have been included to help you understand wildflowers in the context of their environment. How do species adapt to very wet or very dry conditions? What are the special adaptations of carnivorous plants to their habitat? When is fire good for wildflowers? How can a pond become a forest? To fully appreciate the miracle of the natural gardens, these essays will help you see not only what is growing, but why it grows in one particular spot, and how it adapted to do so.

Finally, I provide a natural history of selected plant groups and economic and cultural uses of many species, describing the uses Native Americans and early settlers found for the plants that grew all around them. Fascinating historical data remind us of life in a different time, when people were closer to the earth and the plants (using plants for clothes, dyes, food, and entertainment) and doctored themselves using ancient remedies that were, literally, "tried and true." Once you learn to identify the plant, you can become acquainted with its folklore, adding immeasurably to your enjoyment of the natural world.

I hope that by stimulating interest in these wildflowers, I also stimulate a desire to help preserve the state's natural gardens. Providing descriptions and a knowledge of these plants will allow more eyes to be able to canvass the Lowcountry and Lower Pee Dee in search of rare natural gardens. Somewhere along a river bluff is an undiscovered limestone outcrop with blackstem spleenwort (*Asplenium resiliens,* plate 175), or a swamp edge harboring a few specimens of fever tree (*Pinckneya pubens,* plate 169), or maybe an Indian shell midden, not yet recorded, with a unique assemblage of species. As a botanist who has worked the Lowcountry for thirty years, I still find unique natural areas not yet recorded.

GEOGRAPHIC AREA COVERED

South Carolina is divided into three physiographic provinces: the mountains, piedmont, and coastal plain. The coastal plain, in turn, is divided into two regional belts roughly parallel to the Atlantic Ocean: the inner coastal plain and the outer coastal plain (see map). This book deals primarily with the outer coastal plain, which covers all or parts of fifteen counties. Wildflowers included in the book grow in this outer coastal plain and were photographed in their natural habitat. In addition, the natural gardens described and the sites covered in the guide to selected sites occur in this outer coastal plain.

The area of secondary coverage is the inner coastal plain. With the exception of the eight seaside gardens, examples of the other natural gardens can be found in this area. Indeed, the majority of wildflowers (with the exception of the true seaside species such as sea oats or glasswort) covered in the book also occur in the inner coastal plain.

Although the term Lowcountry is used in the title, no adequate term defines the area. People have defined it geologically, demographically, botanically, and as a "state of mind." Some have defined it as the land from the coast to the inland limits of tidal rice cultivation. Others say that if an imaginary line is drawn connecting the headwaters of the blackwater (coastal) rivers (see map), this line would mark the inland limit of the Lowcountry, thus equating the Lowcountry with the tidal rice-growing area. One may notice that this line roughly corresponds to the inland limit of the outer coastal plain. I feel justified in equating the Lowcountry approximately with the outer coastal plain, with the exception of the Lower Pee Dee.

The Pee Dee area of the state is a major tobacco growing region and includes roughly the eastern side of the state in and around the Great Pee Dee and Little Pee Dee rivers. Four counties, or parts of these counties, occur in the outer coastal plain: Horry, Marion, Florence, and Dillon. This area is considered the Lower Pee Dee and is included in the area of primary coverage since it lies in the outer coastal plain.

For brevity in the text, the term Lowcountry will be used to designate the outer coastal plain, including the Lower Pee Dee.

WILDFLOWERS COVERED

What is meant by the term wildflower? Some sources use it to refer to any species growing without cultivation. Some restrict the term to native, annual, or perennial herbs with showy flowers such as bloodroot and trilliums. However, native trees such as bull bay and tulip tree as well as shrubs like sweet shrub and pawpaw have showy flowers and are more conspicuous than many of the herbs. Many nonnative, naturalized species with showy flowers, such as Japanese honeysuckle, are as showy as the native species and are included in most wildflower books. Many native and naturalized species with small flowers are abundant, aggressive species and often designated as weeds. With magnification, however, their flowers are equally as beautiful as the more showy species. Aggregations of these species along roadsides, as in the case of black medic (*Medicago lupulina*), or in fields, like toadflax (*Linaria canadensis*), give color and character to the land. Certain sedges like marsh bulrush (*Scirpus cyperinus*) and grasses such as giant beard grass (*Erianthus giganteus*) are also equally as conspicuous as the classic wildflowers, so much so that they are often used in dried arrangements. And what plants add more beauty to the Lowcountry than the many woody vines, such as cow-itch (*Campsis radicans*) and coral honeysuckle (*Lonicera sempervirens*)? Some of these are considered noxious weeds, but surely, these are wildflowers.

Many factors were considered when choosing the species to include in this book. The determining factor, however, was to fulfill one of the primary objectives of the book: to ignite people's interest in native wildflower gardens. The showy herbs that abound in these gardens are emphasized; however, selected indicator species of the gardens are included as an aid to identification. For example, turkey oak is shown because it is an indicator species of the xeric sandhills, pond cypress is included to identify the pond cypress savannas, and numerous rushes and grasses are included to identify the brackish marshes along the rivers.

Also, certain species represent interesting accounts about the botanical history of the Lowcountry. Mexican-tea (*Chenopodium ambrosioides*), a weedy introduction, is included because of its use as a folk remedy by plantation slaves; sweet grass (*Muhlenbergia filipes*) is listed because of its past and present use in making sweet grass baskets; poison ivy (*Rhus radicans*) is here because of its poisonous nature; and mistletoe (*Phoradendron serotinum*) is chosen to illustrate a fascinating group of vascular plants, the parasitic species. And many species such as rosemary and American chaff-seed are included because of the recent and growing interest in protection of rare and endangered species.

Another criterion in the choice of species was to include as many species as possible that are not usually depicted in other wildflower books that cover the Lowcountry. Obviously there is still considerable overlap since not to include the common species, even though they are covered in other books, would make this volume a mere supplement. To avoid this, species are divided almost equally between those included in other books and those not.

An additional goal of this book is to assist those who simply want to identify the native and naturalized plants that grow in yards and along roadsides. The essay on the ruderal gardens, which include many naturalized species such as toadflax, common dandelion, and white clover, should fulfill this objective.

Perhaps the final consideration in the choice of which plants to include was to recognize the interest of students and wildflower enthusiasts I have led on field trips over the years. Certain plants, for whatever reason, seemed to attract their attention, whether it was a large-flowered herb or inconspicuous grass or sedge. These species are included in appreciation to them for their interest.

HOW TO USE THIS BOOK

The basic organization of this book is based on my personal concept of how the nonprofessional can identify a wildflower: If one knows its habitat (its natural garden), what the wildflower looks like, and when it flowers or fruits (or has another visual aid such as leaf structure), then identification is made easy to almost anyone who is willing to "go afield."

Getting familiar with the most common natural gardens (technically, plant communities) is the first step. I have chosen twenty-seven natural gardens which harbor the greatest number of wildflowers. Full descriptions of these gardens can be found in part 1, and photographs of each are included in the color plates in part 2. Included in the "Guide to Selected Sites of Wildflower Gardens" in part 3 are locations throughout the Lowcountry where examples of the gardens can be found. In time, the reader will become readily familiar with these gardens.

Grouping wildflowers according to their gardens is a novel approach for a book on wildflower identification. Once readers have identified a particular wildflower garden in the field, they can turn to this garden in the book and scan the color plates for identification of species in bloom or fruit for a particular season. Although some species are restricted to one or two gardens, many species can be found in numerous gardens. The different gardens that each species is found in are given in the species descriptions.

The color plates include photographs of 437 wildflowers grouped according to natural gardens, as well as a photograph of each garden. Within each grouping, the wildflowers are arranged by the months that they are in flower or fruit. Fruits are used instead of the flowers only if they are a better aid to identification. Each wildflower photograph is accompanied by a common name, the scientific name, and the flowering or fruiting month(s). In the few situations where the species is an evergreen, perennial plant (such as live oak, plate 29, or devil-joint, plate 5)—and the entire plant is a better means of field identification—the designation "evergreen" or "all year" is given. Photographs of natural gardens list only the name of the garden. The species descriptions are arranged in the same order as the plates, so the plate number can be used to locate the plant or garden in the "Species Descriptions" section.

The species descriptions contain information on each of the 437 species pictured in the book. The first item listed for each plant is one or several common names designated in the literature and/or from field experience, followed by the scientific name, synonym (if applicable), and the family. Each entry references the plate number where the plant is pictured in the color plates, and contains information about when the plant flowers or fruits.

Next follows a description and range-habitat for each plant, and, if applicable, any taxonomic notes, similar species, or comments. The descriptions are abbreviated versions of descriptions from standard taxonomic texts. Generally, key characters to aid in identification are given that cannot be seen in the color photographs. For example, if the flower is shown, reference is made in the text to the fruits. Or, if the leaves do not appear in the color photograph, pertinent information is given here about leaf structure. Characteristics such as these can be used to confirm or assist in identification. Information that can be gleaned from the photograph is generally omitted.

The range and habitat information is compiled from several sources: available literature, such as the *Manual of the Vascular Flora of the Carolinas* (Radford et al., 1968); my observations from field work in the Lowcountry; records on file with the South Carolina Heritage Trust Program; and communication with other botanists. The information on range is given primarily for South Carolina. The reader may refer to standard manuals for information on range outside the state. The range and habitat data often differ from standard texts because considerable information on the flora of the Lowcountry has been gathered by botanists that has not been published.

The six essays included in "Ecology of the Lowcountry and Lower Pee Dee" are not intended to be a complete coverage of the ecology. They do, however, include material I feel is the most important in determining the diversity of wildflowers that grow in the Lowcountry and Lower Pee Dee. In addition, the essays cover material that the layperson can easily understand. One may refer to the bibliography for references with a more complete account of Lowcountry ecology.

Twelve short essays are included in "Natural History of Selected Groups of Flowering Plants." I have found that a deeper appreciation of wildflowers (and plants in general) is achieved if one knows more than their identification. This "deeper" level comes from a knowledge of such aspects as folklore, ecology, medicinal use, and economic and cultural uses (both past and present) of each species. Most of these essays were developed for *A Field Guide to the Bluff Plantation Wildlife Sanctuary* (Richard Porcher, 1985) and are abbreviated in this book.

Included in the bibliography are works that contain broad information about the ecology and flora of the Lowcountry. Several of these are cited in the text; the remainder are given as sources to augment this text for the reader who wishes to study the Lowcountry in more depth.

PLANT NAMES

Common Names

Common names are not dependable for several reasons: (1) the same plant may have many common names (some 140 common names exist for *Verbascum thapsus*, woolly mullein); (2) the same common name may apply to several plants, either in the same or different areas; (3) they may be misleading: for example, silver-leaved grass is not a grass but a composite; and (4) they are not standardized and are governed by use only, therefore a particular name cannot be judged right or wrong.

In spite of the above problems, common names are often part of the folklore of a plant. Recall the common name "heal-all" for *Prunella vulgaris* in reference to its use in folk medicine as a cure-all, or "boneset" for *Eupatorium perfoliatum* which comes from the belief that it could be used to set broken bones. Common names are often the only means of communication for those unfamiliar with the scientific names. It is also easier to remember words such as "windflower" and "white oak." Often common names are remarkably descriptive of the plant; for example, "blueberry" for a member of the heath family with blue berries, or "swamp chestnut oak" in reference to the leaves of that oak which are similar to those of the American chestnut.

With the above limitations in mind, I have searched the literature, drawn on my knowledge of the plants, and included at least one common name for each plant. Where more than one name is given, the one believed to be more in use in the Lowcountry is listed first.

Scientific Names

Scientific names, unlike common names, are governed by a system of laws called the International Code of Botanical Nomenclature (ICBN). In this system, a plant widely distributed over the world has the same scientific name; furthermore, only one plant can have this scientific name. Also, a plant can have only one valid scientific name. It may have picked up several scientific names along the way, but any names besides the valid one are synonyms.

The scientific (or species) name always consists of two Latin or latinized words (the binomial system of nomenclature), a genus or generic name and a specific epithet, followed by the authority, the name of the person who first described the plant according to the rules prescribed by the ICBN. Given the following scientific name:

Acer rubrum L. or Acer rubrum L.

The generic name is *Acer* and the specific epithet is *rubrum*. The authority is the Swedish botanist Linnaeus; he is honored by letting "L." stand for his name.

Often changes are made in nomenclature. For example, when a species is changed from one genus to another, the name of the original author is placed in parentheses and is followed by the name of the person making the change. The following example illustrates this concept:

Benzoin melissifolia Walter

changed to

Lindera melissifolia (Walter) Blume

Walter originally placed it in the genus *Benzoin;* he now becomes the parenthetical authority. Blume, who changed it to the genus *Lindera,* becomes the authority.

In using scientific names in text, the genus is capitalized while the specific epithet is in lower case. Also, scientific names should always be italicized when printed, or underlined when typed or written by hand. The generic name is sometimes given as an abbreviation if the context makes its meaning clear: for example, *A. rubrum* L., where *A* stands for *Acer.*

Botanists often recognize variations below the species level in plants which are different yet not sufficiently distinctive to be considered separate species. These subdivisions are, in descending order of magnitude, the subspecies, variety, and form. No consensus exists among botanists on the use of the terms subspecies and variety. They are often used interchangeably.

Both of these subdivisions are used in this book, and examples are given here for instructive purposes:

1. Two subspecies of sugar maple, *Acer saccharum* Marshall, are written as:

> *Acer saccharum* subsp. *floridanum* (Chapman) Desmarais

and

> *Acer saccharum* subsp. *nigrum* (Michaux f.) Desmarias

2. Two varieties of *Ilex cassine* L. are given as:

> *Ilex cassine* L. var. *cassine*

and

> *Ilex cassine* var. *myrtifolia* (Walter) Sargent

For the most part, scientific names in this book follow those in the *Manual of the Vascular Flora of the Carolinas.* But for a number of complex reasons, several species are given different scientific names in different texts or manuals. Such synonymous names are unfortunate, but they are part of the vicissitudes of plant nomenclature and have to be recognized. For example, in the manual above, dune spurge is given as *Euphorbia polygonifolia* L., yet in Small's *Manual of the Southeastern Flora* (1933), it is given as *Chamaesyce polygonifolia* (L.) Small. Duncan and Duncan (1987) use this latter form in their *Seaside Plants of the Gulf and Atlantic Coasts.* In order for the reader to correlate these problem species to other manuals, synonyms are included when applicable.

PRESERVATION OF NATIVE WILDFLOWERS

Preservation of our native wildflowers depends on one basic concept that most botanists agree on: plants should be preserved in their natural habitat. A list of wildflowers "not to pick" will never be sufficient. Only large tracts of natural gardens placed under protection will insure future generations the same pleasures we receive when viewing an Easter lily or a pitcher-plant in its natural setting. Today, the South Carolina Heritage Trust Program, the National Audubon Society, the Nature Conservancy of South Carolina, the Lowcountry Open Land Trust, Ducks Unlimited, and other groups have been active in either purchasing land or obtaining conservation easements on private properties throughout the Lowcountry. Both methods insure natural gardens will be afforded continued protection. Many of these protected sites are listed in part 3.

The U.S. Forest Service has also been active in preservation of unique natural areas under its ownership. In the Francis Marion National Forest, studies done in 1980, 1982, 1991, and 1993 identified over fifty-nine natural areas which the Forest Service has agreed to protect and maintain. In addition, four areas in the Francis Marion National Forest have been declared Wilderness Areas, and land around Honey Hill, with hundreds of lime sinks, is being considered as a Research Natural Area.

But much still needs to be done. The areas already protected in the Lowcountry are not sufficient to insure the level of defense necessary to preserve all the species of wildflowers in our area. Several species are known from only a few sites, and an unexpected event such as a Hurricane Hugo could easily eliminate these species.

It takes many eyes to locate unique areas that require protection. One goal of this book is to make people aware of the rare natural gardens in the hope that they will join professional botanists and organizations in first finding, and then protecting, these gardens.

ECONOMIC AND CULTURAL NOTES

Many plants we refer to as wildflowers served the early settlers in several ways. Just as important, they still serve us today. Fuels, building materials, folk remedies and medicines, tools, baskets, and naval stores were made of the native and naturalized plants. Indeed, the history of economic and cultural development of the Lowcountry is laced with plant uses.

The rice industry of South Carolina, which helped shape the culture, economy, social structure, and ecology of the Lowcountry, made use of coiled work-baskets constructed from needle rush (*Juncus roemerianus,* plate 60) bound with thin strips of white oak or strips from the leaves of cabbage palmetto. One of the earliest work-baskets was the "fanner," used to winnow the rice (to throw threshed rice into the air allowing the wind to blow away the chaff). Coiled basketry was probably practiced along the entire range of the rice-growing area (from Cape Fear, North Carolina, to the St. Johns River in Florida). Descended from an ancient African craft, it was introduced into the Carolinas late in the seventeenth century by the African peoples who were brought to America to cultivate rice and other crops.

This art survives today in the Charleston area as sweet grass baskets, not as an agricultural craft, but as an art form. Four plants are used to make the baskets today: needles (leaves) of longleaf pine (*Pinus palustris,* plate 263), stems of sweet grass (*Muhlenbergia filipes,* plate 24), stems of needle rush (*Juncus roemerianus,* plate 60), and strips of the leaves of cabbage palmetto (*Sabal palmetto,* plate 30). Collectable, native populations of sweet grass are becoming scarce or off-limits to the basket makers, threatening the industry. Local people are taking steps to ensure a continued supply of sweet grass for this art.

Grasses also served the settlers in other ways. Salthay (*Spartina patens*) was used in pioneer days as a natural pasture for livestock, while smooth cordgrass (*Spartina alterniflora,* plate 48) was used as thatch for roofs. Sea oats (*Uniola paniculata,* plate 17) and seaside panicum (*Panicum amarum,* plate 23) both continue to build and stabilize dunes along the coast which protect the land from erosion.

Spanish moss (*Tillandsia usneoides,* plate 33) has done more than add beauty to the Lowcountry. Spanish moss fiber was used by early settlers as a binder in construction of mud and clay chimneys and for binding mud or clay in plastering houses. After the Civil War, it was used in overstuffed furniture, upholstery, and mattresses; when the automobile was invented, this flowering "moss" was used for seat stuffing.

Cabbage palmetto yielded a valuable fiber known as "palmetto fiber" obtained from the young leaf stalks still in the bud, while coarser fibers came from mature leaves or the bases of the old leaf stalks. This fiber was used to make various brushes. Palmetto heart (the young, terminal bud) has long been a favorite food of Lowcountry denizens. When cooked, it is a tender, succulent vegetable. This is a wasteful process, however, since cutting out the terminal bud kills the tree. It is mostly gathered now when trees are downed by storms or when land is cleared for construction. Cabbage palmetto first gained popularity when used to make forts in the Revolutionary War; the fibrous trunk would absorb the impact of cannon balls and not shatter. Cabbage palmetto is now the state tree of South Carolina.

Many local gardens are adorned with native shrubs and trees which years ago were transplanted in the dead of winter. Sweet bay, fringe-tree, wild olive, redbud, bull bay, storax, loblolly bay, and beauty-berry, plus many others, can all be seen in gardens around the homes of Lowcountry towns, cities, and villages and in numerous plantation gardens. The native woodlands still provide many vines and herbs that are used to add color to plantation gardens. Vines such as yellow jessamine, climbing hydrangea, cow-itch, cross vine, and coral honeysuckle continue to be used as native landscape materials. Blended with these native plants are a wide variety of introduced species such as crepe myrtle, camellia, azalea, and tea olive.

The flowering trees played an important role in development of the coastal area. Dogwood bark was the source of a drug used as a substitute for quinine to fight malaria. Its wood, resistant to sudden shock, was used to make shuttles in the textile industry which utilized the cotton grown in the Lowcountry. Horse sugar was a source of a yellow dye. Sassafras wood, because of the presence of essential oils that repelled insects (such as bedbugs), was used to make bedsteads and flooring for slave cabins. At one time sassafras oil was considered a miraculous cure-all in Europe and exported from the colonies. Today its oil is used to flavor tobacco, patent medicines, root beer, soaps, and perfumes. Tulip tree is the tallest hardwood in North America and was once one of the most important lumber trees in the East. In the Lowcountry its wood was employed to a great extent by the noted furniture maker Thomas Elf, especially for bedsteads.

Longleaf pine, bald cypress, and cedar were invaluable to the early colonists. Longleaf pine provided tar and pitch, which formed the basis of the naval stores industry. Tar preserved ship's rigging, and pitch was used to caulk wooden sailing vessels. Later, longleaf pine provided spirits of turpentine used in the paint and varnish industry, and resin, which had a variety of uses. Longleaf pine was also the dominant lumber tree in the South, providing "heart pine" boards that grace the floors of many city and plantation homes.

Bald cypress and cedar wood contain essential oils that give the wood natural durability against insects, fungi, and bacteria. These woods were used for shingles and fence posts. Cedar wood was especially popular to build cedar chests as the odor (essential oils) kept away moths and other insects.

Many of the native and naturalized plants of the Lowcountry were used for folk remedies and medicines (see the essay on medicinal plants and folk remedies), while others provided food for the early settlers, as well as for people today.

Ecology of the Lowcountry and Lower Pee Dee

ECOLOGICAL SUCCESSION

Vegetation on most sites is dynamic, not static. Return to an abandoned field after a number of years, and the field of goldenrods and asters has been replaced by a pine forest. Or return to an abandoned rice field along a coastal river in the Lowcountry; what was once a tidal freshwater marsh may now be an immature swamp forest. A process of ecological succession occurs: an orderly sequence of different communities replacing one another over a period of time. If given enough time and no major disturbance occurs, succession eventually terminates in the climax community, that stable end community of succession that is capable of self-perpetuation under prevailing environmental conditions. The entire sequence of communities replacing one another is called the sere and each transitory community in the sere is called a seral stage.

Two general types of succession may occur. One type is primary succession in which a community becomes established on a particular substrate for the first time—that is, no living organisms have previously colonized this particular substrate. In the Lowcountry, succession on newly formed coastal dunes, on sand bars, or on spits along rivers is primary succession. The other type of succession is secondary succession where the substrate has been occupied by vegetation in the past. Some event such as fire, a change in climatic factors, or people's intervention has caused the original community on the site to disappear. Two examples of secondary succession in the Lowcountry are succession in abandoned rice fields and succession in abandoned agricultural lands. In both cases, people have removed the original vegetation (swamp forests where the rice fields were created and upland woods where agricultural fields were created). After abandonment, these areas are today undergoing succession back to swamp forests and woodlands, respectively.

A good example of succession that will help the reader understand this process better is secondary succession in abandoned rice fields along the coastal, freshwater rivers. In abandoned rice fields, hydrarch succession (termed so if succession begins in water) starts with submerged, anchored hydrophytes such as bladderwort (*Utricularia*), waterweed

(*Egeria*), pondweed (*Potamogeton*), and milfoil (*Myriophyllum*). These plants bind the loose soil matrix, trap sediment, and add materially to the accumulation of organic matter as they die. Next, floating-leaved, anchored hydrophytes such as water primrose (*Ludwigia uruguayensis*), alligator-weed (*Alternanthera philoxeroides*), and fragrant water-lily (*Nymphaea odorata*) become established since their submerged stems can reach to the bottom. Some of the floating aquatics, such as water hyacinth (*Eichhornia crassipes*), an introduced species, now become established in the fields. These plants trap more sediment and continue to add to soil accumulation as they die. The submerged, anchored aquatics, their source of sufficient sunlight blocked by the leaves of the floating aquatics, ultimately die. Finally, the soil level has been raised close enough to the surface that emergent, anchored hydrophytes become established. These include persistent emergents such as cat-tails (*Typha* ssp.), marsh bulrush (*Scirpus cyperinus*), rushes (*Juncus* ssp.), giant beard grass (*Erianthus giganteus*), wild rice (*Zizania aquatica*), southern wild rice (*Zizaniopsis miliacea*), and cordgrass (*Spartina cynosuroides*); and the nonpersistent emergents such as swamp rose mallow (*Hibiscus moscheutos*), pickerelweed (*Pontederia cordata*), water hemlock (*Cicuta maculata*), and arrow arum (*Peltandra virginica*).

The emergent species ultimately replace the floating species. In turn these emergent hydrophytes create the conditions for the next seral stage by reducing soil moisture through transpiration, by raising the soil level even higher by trapping more sediment, by decomposition, and by creating a more stable soil by a mass of interlocking rhizomes. A marsh thicket then develops as seeds of woody plants are brought in by wind and/or animals. These include swamp rose (*Rosa palustris*), wax myrtle (*Myrica cerifera*), button-bush (*Cephalanthus occidentalis*), and indigo-bush (*Amorpha fruticosa*). Finally the soil is raised above the water table and the wind-borne fruits of bald cypress, red maple, willows, loblolly pine, cottonwood, sweet-gum, and swamp-gum (*Nyssa biflora*) can become established. This new swamp forest undoubtedly will be different in species composition from the original because many environmental parameters have changed over the years, but it will still be a swamp forest.

For practical purposes the swamp forests that form on the abandoned rice fields are climax communities. There are enough examples along the Lowcountry rivers of swamp forests reaching maturity in abandoned rice fields to determine that they are in relative equilibrium with the environment and are self-perpetuating.

FIRE AND ITS ROLE IN LONGLEAF PINE FORESTS

Many natural communities throughout the world owe their origin and maintenance to fire. These "fire-adapted" communities such as the savannas in Africa, the Douglas fir forests of the northern Rocky Mountains, the grasslands of North America, and the chaparrals of the southwestern United States all have evolved through fire since time immemorial. In the Lowcountry of South Carolina the situation is similar. Longleaf pine savannas (plates 316 and 317) and longleaf pine flatwoods (plate 282) are fire dependant communities that evolved in a fire-dominated area. The pocosins (plate 115) and xeric sandhills (plate 262) also owe much of their character to fires.

Fire has long been a natural part of the Lowcountry. Whenever there was sufficient fuel, dry conditions, and lighting for ignition, fire swept great areas of the Lowcountry—stopped only by natural features such as rivers or heavy rains. Often they swept from river system to river system. When the Indians arrived, they set fires to drive game and clear land around villages. The colonists also practiced the art of using fire as they burned the adjacent woodlands to provide fresh grass for grazing cattle. But year after year in our century, the savannas and flatwoods continue to produce a magnificent display of wildflowers. How can these wildflowers arise every year from the ashes of the forest floor? The answer lies in the special adaptations of the pineland plants.

Fire acts as a selective agent because some plants are more fire resistant than others. Periodic fire burns both herbs, shrubs, and hardwood seedlings and saplings to the ground. Herbs of the longleaf pine flatwoods and savannas have their actively growing parts at or just below the ground surface and are not killed. Shrubs and hardwood seedlings and saplings, on the other hand, have their actively growing parts above ground where they are quickly killed by fire. The longleaf pines are also fire resistant. In the "grass" (seedling) stage, their growing tip is protected from fire by a mass of evergreen needles; as trees, they are fire resistant because of their thick bark.

In the absence of fire, the longleaf savannas and flatwoods quickly change. Shrubs and hardwoods quickly form a dense layer, blocking sunlight to the forest floor and eliminating the herbs. The increased transpiration lowers soil moisture, further restricting herb growth. Ultimately, the beautiful wildflower gardens of the Lowcountry pinelands cease to exist in the absence of periodic fires.

Fire also plays other important roles in the pineland forests. Under an extreme regimen of fire suppression, so much fuel accumulates that control is impossible; either crown fires occur, or the surface fires are so hot that trees that would not normally be damaged are killed. Yearly fires, on the other hand, prevent excessive fuel buildup so that only light surface fires occur and the trees survive. Hot fires also damage soils by destroying the humus layer. Woody debris on the forest floor contains considerable minerals that are released very slowly by bacterial and fungal decomposition; fire releases the minerals immediately. Also, seeds of many trees, including longleaf pine, germinate better on mineral soil free from competition—a condition created by fire.

Ecology

Today "prescribed burning" is practiced by many landowners, timber companies, the U.S. Forest Service, and plantations managed for quail hunting. As a consequence, these lands give us some of our finest flatwoods and savannas for viewing wildflowers. Prescribed burning is setting fire under conditions of the right temperature, humidity, and wind to ensure a light, surface fire. Prescribed burning as a management tool has numerous benefits to pine forests: (1) it prevents litter buildup, reducing the hazards of wildfire; (2) it controls brownspot needle blight; (3) it releases minerals into the soil; (4) it reduces competition from shrubs and hardwoods; (5) it promotes the development of seed-producing legumes important to wildlife; (6) it creates a mineral seedbed for better reproduction of longleaf pine seeds; (7) it promotes growth of the savanna and flatwood wildflowers; and (8) it maintains the longleaf forests in as near a natural state as occurred before the advent of humans in the Lowcountry. Fire, properly used, can be an ecological tool of great value as part of good land management.

RICE CULTURE AND ITS EFFECTS ON THE NATURAL HISTORY OF THE LOWCOUNTRY

The growing of rice from its introduction in 1685 until the end of it as an industry in the early 1900s forever changed the diversity and structure of the flora (and fauna) of the Lowcountry of South Carolina (and similar coastal areas in North Carolina and Georgia). Even today its ecological effects are evident. Plants introduced into plantation gardens have escaped and become naturalized; river swamps that were cleared for rice fields, after being abandoned, today support a variety of plant communities. And upland woods that were cleared for provision fields and abandoned at the end of the industry support a variety of flora and fauna different from the original woods. The rice culture, however, had the greatest influence on the vegetation of the inland and river swamps.

The rice seed that became the principal rice grown was introduced from the island of Madagascar just prior to 1685 and first planted by Dr. Henry Woodward (Salley, 1919). This strain flourished in the swamp lands of the Lowcountry and became known as "Carolina Gold Rice" because of the golden color of the hull. Quickly it became a major export crop of the state and, according to many sources, was the finest strain of rice grown anywhere in the world. It was subsequently introduced into North Carolina and Georgia.

Rice culture quickly influenced every aspect of Lowcountry life. A rich plantation system developed, and the planters, accumulating great wealth, sent their sons to Oxford and Cambridge to be educated. Fine homes were built on the river plantations and were landscaped with formal gardens. Many of the exotic plants that adorned the gardens escaped over time and have become naturalized. Summer homes were built in cities such as Georgetown and Charleston; these homes serve today as tourists attractions, especially in Charleston. A slave trade developed as a source of labor to work the fields; their descendants today are an integral part of the social structure of the Lowcountry.

As malaria became more frequent, the planters settled the inland pinelands and seacoast islands in the summer months (the malaria season). The pinelands were beyond the flight-path of the freshwater swamps where the *Anopheles* mosquitos (the vector of the malaria parasite) breed; also, the sandy soils of the high pinelands did not accumulate pools of water adequate for breeding. Villages such as Plantersville, Pinopolis, and Cordesville sprang up throughout the Lowcountry. Along the coastal islands, where the adjacent marshes breed the salt marsh mosquito, which does not carry the parasite, settlements developed as retreats for the planters. Pawleys Island in Georgetown County, for example, became the retreat for planters of the Waccamaw area.

The industrial base of the Lowcountry began as a result of the rice industry. With the introduction of the horizontal steam engine in the early 1800s, numerous foundries produced engines and accessory equipment to operate the mills that prepared the rice for market (Richard Porcher, 1987). Many companies today that serve the Lowcountry are descendants of these early foundries.

More pertinent to this book, however, is the effect of the rice industry on the natural history of the Lowcountry. Rice growing in the Lowcountry can be separated into two distinct periods: the inland swamp and tidal swamp systems.

The inland swamp system, the initial method of growing rice, began around 1700 and lasted until the end of the American Revolution when it was replaced by the tidal system. The inland swamp system depended on rainwater to fill reservoirs that supplied water for the crop. An earthen bank was thrown up across the upper part of an inland swamp, from highland to highland. In their upper reaches the inland swamps supported freshwater swamp forests while in their lower reaches, where they drained into coastal streams or rivers, they supported fresh or brackish marshes. Further down from the upper bank a second bank was constructed similar to the upper. The area between the two banks was then cleared of trees, ditched, and made ready for planting. Water from the reservoir, through a trunk-gate system, could then be applied to the fields. Another trunk-gate system in the lower bank was then used to drain the water from the fields when the tide was low in the adjacent stream. With a system established for controlling the source of water, sufficient crops could be grown to make it profitable. Two excellent examples of inland rice fields occur on the Bluff Plantation Wildlife Sanctuary in Berkeley County and Tea Farm County Park in Charleston County.

The inland system had one main drawback in producing a marketable crop: the reservoirs depended on rainwater. A drought, which often occurred, meant loss of the crop. It is understandable, then, that a more certain water supply for rice growing was found: fresh water from the tidal rivers that traverse the Lowcountry.

These freshwater rivers were bordered by fringes of freshwater tidal marshes (plate 65) and wide expanses of low-lying, freshwater tidal swamps (plate 96). Twice a day, as far as 30 miles inland, the tide ebbed and flowed, draining then flooding the marshes and swamps. Beginning around the middle 1700s, an ingenious system was devised to apply this "rhythm of nature" to rice growing, thus ensuring a consistent supply of fresh water. On every river in the Lowcountry above the influence of salt water and up-river to the point where at least a 3-foot difference in low and high tide occurred, an attempt was made to grow rice. A bank was constructed from the highland through the swamp to the river's edge, then along the river's edge, then back through the swamp to the highland. This bank kept the river water out of the area during high tide. A series of lower "check-banks" were constructed within the large area, dividing it into smaller fields. Each field was fitted with a trunk-gate system so each could be flooded or drained independent of the other fields. Next the task of clearing the swamp began. Slaves, using primitive hand tools and oxen, felled, piled, and burned the trees. The largest trees were cut at ground level; their stumps can still be seen today in the abandoned fields. Fields were then made ready for planting. Using the trunk-gate system, each field could be flooded when the tide rose in the river, or drained at low tide. This dependable supply of fresh water made rice growing more profitable. By the end of the American Revolution, the tidal system replaced the inland system as the principal method of growing commercial rice, and remained the basic method of commercial rice culture until the end of the industry in the early 1900s.

Both systems of growing rice had major effects on the natural history of the Lowcountry. Many acres of inland swamps were cleared for fields. When the inland system was abandoned, the fields reverted back to swamp forests. No records exist as to the species composition of these original inland swamps; thus, it is not known how similar the present swamp forests are to the original. But the most pronounced legacy of the inland system is the reservoirs that still exist. In many cases the upper bank is still intact, creating a permanently flooded swamp that provides valuable habitat for many species of animals, especially for wading birds. Washo Reserve on the Santee Coastal Preserve, the Reserve Swamp at the Bluff Plantation Wildlife Sanctuary and Cypress Gardens represent just three of the former inland reservoirs.

The tidal system has also left a major legacy: approximately 150,000 acres of abandoned tidal fields along the rivers, many of which still have their banks intact and are managed for waterfowl. The ability to control the water in the fields allows a management regime to select plants preferred by waterfowl, such as redroot (*Lachnanthes caroliniana,* plate 353) or widgeon grass (*Ruppia maritima*). Other fields with broken banks that allow the tides free access are undergoing a successional process that will result in swamp forests reclaiming the fields. In fact, many of the fields have already reverted to swamp forests. Once again, whether the mature secondary swamp forests will be similar to the original is not known. It is in these abandoned tidal fields that one of the greatest displays of wildflowers occurs: those of the freshwater tidal marshes. One can paddle a small boat through the broken banks and follow a myriad of canals, each revealing a different combination of freshwater marsh plants in the spring, summer, and fall.

Rice growing generally ended in the Lowcountry around 1910–1911 when two disastrous hurricanes struck two years in a row—breaking the banks, flooding the fields with salt water, and destroying the crops. These storms, combined with the loss of slave labor after the Civil War and the advent of rice growing in Arkansas, Texas, and Louisiana around 1880 (where rice growing was mechanized), made it impossible to grow rice profitably in South Carolina.

After the Civil War, and after the demise of the industry, many plantations were bought by wealthy northerners who used them as hunting preserves or retreats from business pressures. Even though there was much resentment in the South at having had to lose their lands to "outsiders," it was a fortuitous event since many of these plantations have been established as wildlife preserves. Some northerners came to love the land so much that, rather than see these plantations developed to serve a select few, they took steps to preserve them forever for the people of South Carolina. A prime example: Tom Yawkey, who owned the Boston Red Sox, gave the state 20,000 acres on the North Santee River. Today it is managed by the South Carolina Department of Natural Resources as the Tom Yawkey Wildlife Center. Other such gifts include the Bluff Plantation Wildlife Sanctuary, the Santee Coastal Preserve, and Hobcaw Barony.

CAROLINA BAYS

Carolina bays (figure 1) are geological formations of unknown origin that occur mainly on the South Atlantic Coastal Plain in North and South Carolina, and Georgia. They are shallow depressions in the sandy coastal soil that vary in depth from a few feet to around 20 feet. They vary in length from a few hundred yards to several miles. The bays occur in three shapes: elliptical, oval, or asymmetrical, with their long axis oriented in a southeast direction. Often the bays overlap each other, as seen in outlines of smaller bays in the depressions of larger ones. An estimated one hundred thousand occur in the three states. Furthermore, their distribution is nonrandom as they occur in clusters; within the clusters they are often aligned, apparently along some undetermined, physiographic gradient.

Although for years several workers had made references to "certain depressions" in the coastal plain, it was not until 1933

Figure 1.
Little Ocean Bay
in Berkeley County

when two scientists, Melton and Schriever, viewed aerial photographs of terrain near Myrtle Beach, South Carolina, that the true characteristics of the bays became known. The photographs revealed sand ridges that are more prominent on the southeast side of the bays. Melton and Schriever then came to an astonishing conclusion: the depressions were formed by a shower of meteorites coming from the northwest at an angle of 35–55 degrees. The impact of the meteorite formed the depression and pushed up the sand ridge.

Their theory was viewed with skepticism by some and accepted as dogma by others. The scientific community has never accepted an extra-terrestrial (or meteoritic) origin of the bays since so much convincing evidence against it has been gathered. At the same time, many laypersons will accept no other theory. As recently as 1982, Henry Savage, in his book *The Mysterious Carolina Bays,* argues for the meteorite theory of origin. His book, a must for students of the bays, is the first in recent times to present a complete review of the subject and includes an extensive bibliography.

Of all the theories that have been postulated by scientists to explain the origin of the bays, two general themes have been dominant: (1) the catastrophic, which envisioned the sudden shower of meteorites (extra-terrestrial), or the sudden formation of artesian springs or sink holes (terrestrial); and (2) the uniformitarianist, which says the origin must be explained by gradual effects of wind, excavation, soil solution, and wave-induced erosion.

But why is there still such a controversy over the origin of the bays with all the theories presented by the scientific community? Simply this: no theory presented adequately explains all the observed facts surrounding the bays. Until one does, their origin will remain surrounded by uncertainty and mystery.

Whether or not the riddle of their origin is ever solved, the

Carolina bays afford a wildflower paradise. In the deeper bays, swamp forests develop. If cypress is a component, it is always pond cypress (*Taxodium ascendens*), either in pure stands or mixed with swamp-gum. In more shallow bays, the beautiful pond cypress savannas (plate 370) may grow. For the most part, however, the Carolina bays harbor pocosins (plate 115). Here occur the three bay trees: loblolly bay, sweet bay, and red bay. Did the Carolina bays get their name from the presence of the bay trees? Or did the bay trees get their names from occurring in the bays? (The depressions were called bays in earlier times.) Again, we have another mystery. The sand ridges are also an important component of the bays. Here the deep sands harbor the xeric sandhills community (plate 262) which contains many interesting and beautiful wildflowers.

Ecologically the bays are important for several reasons. First, they are wetland habitats and are an oasis for numerous animals of the surrounding uplands. Two endangered animals, the black bear and pine barrens tree frog, make their homes in the pocosins. Venus' fly trap (*Dionaea muscipula*, plates 334 and 335), and lamb-kill (*Kalmia angustifolia* var. *caroliniana*, plate 124), two uncommon species, grow on the margins of the pocosins. Add the savanna gardens with their floral display of orchids and carnivorous plants and the xeric sandhills of the ridges, and one can easily see why the Carolina bays are such important habitats.

Four areas in particular harbor a variety of Carolina bays protected for the public use: Lewis Ocean Bay Heritage Preserve and Cartwheel Bay Heritage Preserve in Horry County, Santee Coastal Preserve in Charleston County, and the Francis Marion National Forest in Berkeley and Charleston counties.

Unfortunately, the majority of the Carolina bays are highly disturbed. Many have been drained for farmland or timber production, others have been converted to pastures, and still others have been used as junk yards to dispose of household goods. Saving the bays has become a major goal of the Nature Conservancy of South Carolina and the South Carolina Heritage Trust Program; several bays have been placed under their protection for future generations. Hopefully, more will be saved.

PLANT ADAPTATIONS

The myriad of ways in which plants have adapted to their environment has always been of special interest to ecologists. Any feature of an organism or its parts which is of definite value in allowing that organism to exist under the conditions of its habitat may be called an "adaptation." Many of the adaptations are so subtle that most observers are never aware of them although they may be familiar with the plant. Others adaptations, however, are obvious and are understood by casual observers. Often these adaptations are not ecologically understood or are enveloped in misconceptions. How many people in the Lowcountry believe that Spanish moss is a parasite and not an epiphyte, or that Jack-in-the-pulpit is a carnivorous plant because of its spathe which resembles the pitcher-plant leaf? Below, some of the salient and subtle ways plants have adapted to environments in the Lowcountry will be discussed.

Xerophytes are plants that live under extreme water stress—that is, they live in habitats where it is difficult to obtain fresh water. Halophytes are xerophytes that live in a saltwater environment and are generally defined as those with a tolerance of 0.5 percent NaCl in the soil. The salt in the water, not the lack of fresh water, makes it difficult for halophytes to absorb fresh water from the environment. Hence, they are often called "wet" xerophytes to distinguish them from the true "dry" xerophytes. Numerous adaptations evolved to allow these plants to derive their fresh water from salt water and for conserving the fresh water they obtain. One such adaptation is the presence of salt glands that excrete the excess salts dissolved in soil water. One can see or taste the excreted salt on the leaves of smooth cordgrass. These marsh grasses also have a thick and continuous cuticle and specialized cells called bulliform cells which collapse when they lose water, resulting in a lengthwise curling of the leaf. This curling of the leaf reduces the surface area exposed to the atmosphere, thus reducing water loss through the stoma. Another adaptation is fleshy tissue (succulence)—the increase in size or abundance of certain internal cells that provides additional storage space for water. Three examples of plants with fleshy tissue that occur in the salt marshes of the Lowcountry are saltwort (*Batis maritima,* plate 53), sea purslane (*Sesuvium portulacastrum,* plate 11), and perennial glasswort (*Salicornia virginica,* plate 52).

The true or "dry" xerophytes are defined as plants growing on substrata that usually become depleted of soil water to a depth of at least 8 inches during a normal season. Conditions such as this occur in the Lowcountry on the sand ridges bordering Carolina bays and along the fluvial ridges of rivers. On these ridges the sandy soil has little water-holding capacity resulting in xeric conditions even though rainfall is abundant. Numerous adaptations have evolved to allow plants to survive in the xeric sands. Sandhills baptisia (*Baptisia cinerea,* plate 270) and sandhills milkweed (*Asclepias humistrata,* plate 271) orient their leaves in a vertical manner, reducing the impact of sunlight and heat from soil reflection during the time of day when light and heat are most intense. Succulence is also exhibited by the xeric sand species as a way of retaining moisture, like sandhills milkweed and prickly-pear (*Opuntia compressa,* plate 259). Some species have narrow leaves that reduce the amount of water loss. Wire plant (*Stipulicida setacea*) carries out its vegetative growth in early spring when moisture conditions are more favorable and spends the hot months from July to September and the succeeding winter in the seed stage.

Dunes along the coast also present an inhospitable place for plants as the shifting sand alternately covers and uncovers the dune plants; however, the plants are the principal agents contributing to establishment and maintenance of the dune system by creating windbreaks that cause the windblown sand to settle-out, thus building a dune. As the plants become buried, they have the ability to produce new growth, creating additional windbreak potential. Their extensive root and rhizome systems, produced in the above process, stabilize the dunes. The primary species on the dunes along the Lowcountry coast are four grasses: sea oats (*Uniola paniculata,* plate 17), seaside panicum (*Panicum amarum,* plate 23), dune sandbur (*Cenchrus tribuloides,* plate 22) and saltgrass (*Distichlis spicata*).

Other adaptations evolved in plants growing on the dunes as a result of exposure to saltspray. The diminutive forms of camphorweed (*Heterotheca subaxillaris,* plate 18) and horseweed (*Erigeron canadensis,* plate 19) reduce surface area that would be exposed to the killing effects of saltspray; beach morning-glory (*Ipomoea stolonifera,* plate 20) has a low, trailing growth that keeps it from the full force of the saltspray.

Hydrophytes include aquatic plants which normally grow in water or inhabit soils containing an amount of water that would prove suboptimal for the average plant. One of the most outstanding structural features shared by most hydrophytes is aerenchyma tissue, formed by the disintegration of groups of cells, or the separation of cells, creating enlarged, intercellular cavities (lacunae) that become filled with gases. Two groups of hydrophytes, the floating and the floating-leaved anchored, have leaves that float on the water's surface because of these gas-filled cavities. Water hyacinth (*Eichhornia crassipes,* plate 109), a naturalized species, is a floating aquatic plant with no connection to the soil, while fragrant water-lily (*Nymphaea odorata,* plate 111), cow-lily (*Nuphar luteum,* plate 106), water-shield (*Brasenia schreberi,* plate 114), pondweed (*Potamogeton pulcher,* plate 112), floating bladderwort (*Utricularia inflata,* plate 108), and frog's-bit (*Limnobium spongia,* plate 113) are floating-leaved, anchored hydrophytes. Both groups have mechanisms that render the upper leaf surface difficult to wet. In fragrant water-lily a waxy surface causes water drops to roll off quickly, thus allowing exchange of gases to occur through the stoma located on the upper leaf surface. The fruits of cow-lily and sacred bean also have large air cavities enabling them to float, which aids in dissemination.

Pocosins generally occur on flat uplands which are water-logged for long periods when the rains are frequent, but may be dry during extended droughts. During the water-logged period, little or no oxygen is available in the soil, restricting growth and maintenance of the root systems of the shrubs. Thus, water intake by the roots is limited, and unless water loss by the leaves is reduced, the plant comes under extreme stress and may die. One method to reduce transpiration from the leaves is a thick, waxy cuticle, and the pocosin shrubs exhibit this to the extreme. One only has to observe the leaves of the following characteristic pocosin species to understand this concept: fetterbush (*Lyonia lucida,* plate 125), bamboo-vine (*Smilax laurifolia,* plate 134), and honeycup (*Zenobia pulverulenta,* plate 126). Another feature of the pocosin shrubs is their ability to resprout from their base following fire. The frequent fires that sweep the pocosins burn everything above ground; their bases, however, survive in the damp peat and very quickly regenerate the above-ground growth. Six months after a fire, the pocosin appears as green and lush as it was before a fire. Interestingly, most bay shrubs also contain volatile compounds that burn readily and actually promote fire.

In this pocosin habitat, pond pine (*Pinus serotina,* plate 116) adapts to the periodic fires that sweep the pocosins during dry spells. Pond pine has latent axillary buds hidden under the bark of the stem and at the base of the tree that facilitate shoot replacement after a crown fire has killed the above-ground branches. Heat triggers a hormonal mechanism that causes the buds to break dormancy and grow rapidly into new shoots. One can readily observe these trees in the Lowcountry pocosins. They are recognized by having short, dead branches on the main trunk, with a series of dense, green tufts (new shoots) scattered along the trunk. The term serotinus, meaning "late to open," refers to another fire adaptation of pond pine. Some of the cones on a tree remain unopened with viable seeds for years. Such cones, when fire spreads through the pocosin, gradually open as heat softens the resinous seal of the bracts which allow the cone to dry out. This expansion then creates internal stresses that force the scales open. The winged seeds flutter to the ground where they find an exposed mineral soil free from competition.

Species of *Kalmia* have evolved a mechanism to facilitate cross-pollination. Their stamens are reflexed and held in pockets of the petals. When an insect scrambles over the flower, it jars the stamens from the pockets, allowing them to spring inward, dusting pollen on the insect. When the insect visits another flower, the pollen brushes onto the stigma, thus effecting cross-pollination. One can observe this feature in lamb-kill that occurs along the edges of pocosins and hairy wicky in the pine flatwoods.

In the longleaf pine savannas a crafty scheme evolved in the orchid genus *Calopogon* that deceives pollinators. These orchids have a tuft of yellow, clubbed bristles (plate 320) on their erect lip which resembles (mimics) the yellow stamens of smooth meadow-beauty (*Rhexia alifanus,* plate 343) which the insect visits for nectar. The orchid has no nectar in its flowers. Occasionally, then, when an insect pollinator mistakes the orchid for the meadow-beauty and picks up pollen from the orchid, cross-pollination is effected if the insect is again fooled into visiting another orchid flower.

The trunk of the majestic bald cypress (*Taxodium distichum,* plate 135), when growing in saturated or seasonally submerged soil, produces an enlarged base, the buttress, which has important survival value. The wide buttress gives the tree a broad base of support in the soft, swamp soil. Indeed, it is seldom that a bald cypress is blown down. Other swamp trees, such as tupelo gum, swamp-gum, and pond cypress also form buttresses.

MARSHES, SWAMPS, AND PEATLANDS

Marshes, swamps, and peatlands are wetlands. There is no single, accepted term in the scientific community to define a wetland. In general terms, however, wetlands are lands where saturation with water is the dominant factor determining the nature of soil development and the types of plant and animal communities living in the soil and on its surface.

Marshes are wetlands inhabited by herbaceous plants rooted in the substrate but with photosynthetic and reproductive organs principally emersed. The dominant species are grasses, rushes, and sedges, along with numerous broadleaf, flowering plants. Swamps, on the other hand, are wetlands dominated by woody plants, where the substrate is flooded for one or more longish periods during each year, sometimes more or less permanently flooded, but usually without surface water part of the time. Both marshes and swamps have predominantly mineral (nonorganic) soils, even though much organic material may be incorporated. Peatlands, simply, are wetlands whose soils are peat, the partially decomposed remains of dead plants and, to a lesser extent, animals.

Marshes

Marshes of the Lowcountry can be classified as either salt, brackish, or fresh. Salt marshes generally grow on a peaty substrate along tidal inlets and behind barrier islands and spits. They are regularly flooded and drained by tidal action; the soil is saturated, high in salts, and low in oxygen. The dominant salt marsh in the coastal area is the *Spartina* marsh vegetated by smooth cordgrass (*Spartina alterniflora,* plate 48). Vast stretches of this marsh system are readily visible as one drives from the mainland to the barrier islands such as Kiawah, Folly, Hilton Head, and Pawleys.

Brackish marshes (plate 57) are transitional communities between the freshwater and salt marshes. They occur where fresh water and salt water mix so that some species of both systems occur together, creating the transitional community. The dominant species are emergent grasses, sedges, and rushes. The brackish marshes occur at varying distances up-river from the coast. The more fresh water coming down the river, the closer the brackish zone occurs to the coast.

Freshwater marshes occur both inland and along the coastal rivers. Inland freshwater marshes are a diverse system, fed by inflowing water, seepage, and precipitation. The variable water supply results in flooding during high rainfall and drawdown during dry periods—a feature that shapes the structure and composition of the marsh. Inland marshes occur along the edge of lakes and ponds, in canals and roadside ditches, and along the edge of inland swamps.

The dominant freshwater marshes of the Lowcountry, however, are the tidal freshwater marshes (plate 65). These marshes occur along the edge of the brownwater and blackwater rivers and are close enough to be affected by daily tides but distant enough to be unaffected by intrusion of salt water. Vast expanses of these marshes occur in the abandoned rice fields along the rivers; narrow zones also occur where the tidal swamps border the rivers. The low marsh with its deeper water is characterized by broadleaf monocots such as arrow arum and pickerelweed and showy dicots such as bur-marigold. The high marsh is a mixture of low marsh species plus numerous grasses, rushes, and sedges.

Swamps

The swamps of the Lowcountry are extensive and varied and include the bottomland hardwood swamps, tidal freshwater swamps, and nonalluvial swamps. The bottomland hardwood swamps occur on the alluvial floodplains of the brownwater and blackwater rivers. The brownwater rivers originate in the piedmont and mountain areas and have wide, alluvial floodplains; the blackwater rivers originate in the coastal region and have narrower, less-developed, alluvial floodplains. On these floodplains occur two bottomland swamps. On the floodplains where the land is almost continuously flooded, the bald cypress–tupelo gum swamps occur (plate 135). Here the trees exhibit typical hydromorphic features in response to growing in water: buttresses, knees, and spongy roots. Hardwood bottoms (plate 162), on the other hand, occur on floodplains slightly elevated above the adjoining swamp forests. Here the land is often flooded, but is dry through much of the year. The dominant vegetation is a mixture of water-tolerant, deciduous hardwoods with oaks predominating, and an occasional cypress from the adjacent swamp forests.

Tidal freshwater swamps (plate 96) occur from the upper limit of tidal influence to the brackish water line downstream. Tidal freshwater swamps owe their nature to the river tides: twice a day they are flooded and twice a day they are free of surface water. Both brownwater and blackwater rivers of the Lowcountry support tidal swamp forests.

A variety of nonalluvial swamps occurs in the Lowcountry. The main feature that distinguishes these swamps is that they are not associated with moving river water; in other words, they are not alluvial systems as are the bottomland hardwoods. These swamps occur in two types of habitats: (1) upland, isolated sites such as lime sinks, Carolina bays, and irregular depressions, and (2) seepage slopes associated with river systems along the boundary of floodplains and uplands. The water supply for the bays, sinks, irregular depressions, and seepage slopes is predominately rainwater. In seepage areas, rainwater that hits the uplands moves downward in the soil until it hits a confining layer. It then moves laterally until it emerges as seepage at the base of the slope.

The swamps of the upland sites are dominated by pond cypress (*Taxodium ascendens*) and swamp-gum (*Nyssa biflora*), either in pure stands or mixed. In the lime sinks, which are low mineral sites, few shrubs or herbs are present (plate 382). The seepage slope swamp forests are diverse with flora similar to the swamp forests of the adjacent floodplain; however, swamp-gum replaces tupelo gum (*Nyssa aquatica*) as the dominant gum.

Peatlands

The dominant peatlands in the Lowcountry are the pocosins, or evergreen shrub bogs. Pocosins (plate 115) are freshwater wetlands found extensively in the lower coastal plains of Virginia, North and South Carolina, and Georgia. They are fire-adapted systems dominated by a tangled mass of broadleaf evergreen or semi-evergreen shrubs and vines with emergent, scattered pond pines and occasional bay trees (red bay, sweet bay, and loblolly bay). The most extensive pocosins are found on flat upland areas; additionally, pocosins may be found in Carolina bays, wet seepage slopes, and in low areas of relict dune fields.

Descriptions of Wildflower Gardens

Lowcountry wildflower gardens are here for all to see and appreciate. One need only spend time in the field, either in the proximity of home or further off the beaten path, to be awakened to their beauty. From the dunes along the coast to the swamps and marshes of the coastal rivers, and inland to the deciduous forests or to the great pine forests, a wide variety of wildflower gardens exists—both natural and ruderal—with each harboring its own display of wildflowers. One can venture forth in search of wildflowers during any season. This section will introduce the reader to the different natural and ruderal gardens of the Lowcountry.

The natural gardens have been described by various workers, either in reference to the Lowcountry only or as part of the natural vegetation of the entire state. The following account has been gleaned from sources included in the bibliography and augmented by my field observations. A photograph of each garden can be found in the color plates. The "Guide to Selected Sites of Wildflower Gardens" in part 3 contains directions to forty-eight sites where examples of the following gardens can be found; the table of sites can be used for an easy cross-reference.

THE SEASIDE GARDENS

Coastal Beaches

Along the South Carolina coast, beaches (plate 1) form where ocean currents and waves deposit sand picked up by the waters from offshore coastal sites or brought down the rivers from inland areas. No vascular plants grow along the beach below the hightide line because it is a dynamic zone, constantly changing with the action of the wind and tides. Above the zone of highest tide, a zone of detritus (driftline) develops which is deposited by the tides. The dominant component of the detritus is the remains of smooth cordgrass from the nearby salt marshes that washes down the tidal creeks to the ocean. Seeds of two hardy species, sea rocket (*Cakile harperi,* plate 2) and Russian thistle (*Salsola kali,* plate 3), are generally the first species to become established in this harsh environment. Both are cosmopolitan waifs—being pioneer species in numerous harsh environments worldwide.

Coastal Dunes

Landward from the driftline is the berm—a zone of fairly level, loose sand subject to heavy saltspray, and above all but the highest spring tides. Here the coastal dunes form (plate 4). Dunes are mounds of unconsolidated sand formed in the berm area by winds blowing across the beach. Whenever an object reduces the force of the wind, the sand it carries is dropped, forming the dune. Coastal dunes front the barrier beaches and barrier islands of the Carolina coast, interrupted periodically where inlets and sounds allow the ocean to surge through dissipating its energy in the tidal flats and marshes inward. Dunes are also the first line of defense for the interior against oceanic forces, especially hurricanes and winter storms. Just as winds can build dunes, they can destroy what they have created. At a point where the vegetation has been killed, the wind can erode dunes. Preservation-minded coastal residents have long fought to protect dunes from excess disturbance such as trails made by people or dune-buggies; others, however, see the coastal area in light of short-term profits with little regard to the future. The South Carolina law against disturbing sea oats on public property has done much to protect the coastal dunes.

The plant most associated with building coastal dunes in the southeast is sea oats (*Uniola paniculata,* plate 17). B. W. Wells, in his *Natural Gardens of North Carolina,* described the role of sea oats in building the dunes:

> On the sand-flat just above high tide, a seedling sea oats plant becomes established. When its leaves extend above the surface, a small windbreak is formed, with the immediate result that a tiny mound of sand accumulates behind and against the little leaf cluster. This additional bit of soil makes possible further growth of the grass, which improves the windbreak; more sand piles up, more grass grows, and in a few years a dune many feet high is built.

Other grasses that help build the dunes are seaside panicum (*Panicum amarum*) and dune sandbur (*Cenchrus tribuloides*). Once the dunes become fairly stable, various forbs become established and help further stabilize the dunes: beach pea (*Strophostyles helvola*); horseweed (*Erigeron canadensis* var. *pusillus*); two primroses: beach evening-primrose (*Oenothera drummondii*) and dunes evening-primrose (*O. humifusa*); camphorweed (*Heterotheca subaxillaris*); seaside pennywort (*Hydrocotyle bonariensis*); and two species of cactus: prickly-pear (*Opuntia compressa*) and devil-joint (*O. drummondii*), among others.

Between the dunes develop the swales (plate 4), low-lying areas protected from the salt-laden winds where a fresh to brackish system may develop because of rainwater collecting and forming a system above the heavier salt water. In this microhabitat common marsh-pink (*Sabatia stellaris,* plate 16) is especially abundant.

On barrier islands or barrier beaches where accretion is occurring and the dunes are building seaward so that the inner dunes become protected from the salt-laden winds, or where the shoreline has been stabilized for years, the dune communities may be replaced by maritime shrub thickets, which may in turn be replaced by maritime forests.

Maritime Shrub Thickets

Maritime shrub thickets (plate 25) occur in the swales between and on the tops of stabilized dunes. Unlike the salt shrub thickets, which occur landward of the maritime forests, the maritime shrub thickets occur seaward of the maritime forests. It is not flooded by salt water except during hurricanes. On top of the dunes, a pronounced stunted, shrubby growth occurs due to the effects of the salt-laden winds that sweep the dune tops. Dominant plants are shrubs and vines. Shrubs include *Myrica cerifera* (most common and often forming almost pure stands), *Baccharis halimifolia,* and *Ilex vomitoria.* Vines include dune greenbrier (*Smilax auriculata,* plate 27), poison ivy, Virginia creeper, Japanese honeysuckle, and supplejack.

On barrier beaches where accretion is occurring and the dunes become more removed from the ocean's edge and the effects of saltspray, maritime forests replace the maritime shrub thickets as live oak and other maritime forest trees become established.

Maritime Forests

Maritime forests (plate 28) occupy the barrier islands and barrier shores of the South Carolina coast. The characteristic species are a variety of salt-tolerant, evergreen trees and shrubs. There is little herbaceous cover except in exposed sites due to natural breaks in the canopy or in disturbed areas.

On the ocean side, maritime forests are shaped (literally) by the effects of saltspray. The wind blowing from the sea carries salt, depositing it on the windward branches and leaves. These leaves and branches die from the salt, while those on the leeward side, protected from the saltspray, continue growing. The result is a "shearing effect" of the trees and shrubs. Inland, away from the effects of saltspray, the trees assume a more typical appearance.

The characteristic evergreen trees of maritime forests are live oak (*Quercus virginiana,* plate 29), bull bay (*Magnolia grandiflora,* plate 37), loblolly pine (*Pinus taeda*), and laurel oak (*Quercus laurifolia*). The subcanopy includes the evergreen trees cabbage palmetto, American holly (*Ilex opaca*), red bay (*Persea borbonia*), and Hercules'-club (*Zanthoxylum clava-herculis,* plate 31). In more open sites wax myrtle, yaupon, wild olive (*Osmanthus americana,* plate 34), and southern red cedar occur. Herbaceous wildflowers are sparse because of the dense canopy of trees, but two species can be found in more open sites: prickly-pear cactus and trailing bluet (*Houstonia procumbens,* plate 36).

The maritime forests along the Atlantic Coast have been extensively timbered since colonial times. No original growth stands exist along the South Carolina coast with the possible exception of the interior of St. Phillips Island in Beaufort County where low sloughs prevented timbering. In fact, islands such as Capers Island in Charleston County and Daufuskie Island in Beaufort County had extensive agricultural fields in the interior; their present forests are secondary growth.

Timbering began in the maritime forests in the 1700s, mainly for live oak to build wooden sailing vessels. After the War of 1812, "live oak mania" began as expeditions were sent from the northern shipyards to the Atlantic and Gulf coasts to harvest live oak. After the invention of iron and steel ships, live oak was given a reprieve; timber companies then began to harvest the pines of the maritime forests. The maritime forests of the Lowcountry today are secondary forest; the large live oaks are either ones that were left for whatever reasons, or that came from seedlings. Live oak is a fast growing tree on good sites and can reach impressive size in fifty years. One only has to understand that the large live oaks that line plantation avenues were planted starting in the 1700s and note their large size today.

Salt Shrub Thickets

Salt shrub thickets (plate 41) occur in a narrow zone between the salt marshes and maritime forests. The two most indicative species are sea ox-eye (*Borrichia frutescens,* plate 42) and marsh-elder (*Iva frutescens,* plate 45). The distinct zone exists because of the differential degree of tidal flooding; the lowest zone floods only at extreme high tides, the highest zone only during storms or hurricanes. Needle rush (*Juncus roemerianus,* plate 60) generally forms a dense growth at the lowest zone. Above the needle rush is a zone of perennial glasswort, sea lavender, and salthay (*Spartina patens*). The highest zone is a shrub border of sea ox-eye, marsh-elder, sea myrtle (*Baccharis halimifolia*), and yaupon (*Ilex vomitoria*). Mixed in with the shrubs, or landward to them, is a variety of trees including cabbage palmetto and southern red cedar. Other species may include wax myrtle (*Myrica cerifera*), sand-vine (*Cynanchum palustre,* plate 43), seaside goldenrod, salthay, and broomstraw (*Andropogon* sp.).

Salt Marshes

Salt marshes (plate 48) occur on regularly flooded, peaty substrates along tidal inlets and behind barrier islands and spits. This community is species-poor, often supporting pure stands of smooth cordgrass (*Spartina alterniflora*); however, it is one of the most productive in the world. It is the dominant wetland in the coastal zone of South Carolina, comprising approximately 150,000 acres. As the cordgrass dies, it is washed into the inlets and ocean where it forms the basis of the estuarine food chain.

Three forms of smooth cordgrass occur: tall, medium, and

short. The tall *Spartina* grows next to the tidal creeks in relatively deep water where it receives an energy subsidy. Away from the creek the tall cordgrass grades into the medium, then the medium grades into the short which occurs at the highest elevation where it is flooded daily, but only to a depth of a few inches to a foot or so.

In the short zone, two species that also occur in the adjacent salt shrub thickets intermix with the short *Spartina:* sea lavender (*Limonium carolinianum,* plate 49) and saltmarsh aster (*Aster tenuifolius,* plate 50), adding a touch of color to the otherwise drab salt marshes.

Salt Flats

Salt flats (plate 51) are formed where tidal waters drain away incompletely and the soil becomes hypersaline as the water evaporates, leaving behind the salt which often forms a white crust on the soil. Even the most salt-tolerant species cannot survive in these hypersaline areas, and the center of the flats are often barren. However, as salinity decreases toward the margins, a variety of fleshy halophytes, salt-loving grasses, and other herbs appear. Closest to the center are perennial glasswort (*Salicornia virginica,* plate 52) and saltwort (*Batis maritima,* plate 53), the two species that tolerate the highest salinity. Salt flats grade into either salt marshes or salt shrub thickets. Intermixed with saltwort and glasswort are diminutive forms of species associated with these latter gardens: sea ox-eye, marsh-elder, smooth cordgrass, sea lavender, saltmarsh aster, and seaside goldenrod (*Solidago sempervirens*).

Shell Mounds

Scattered along the coast in the salt marshes and at the tips of land masses within estuaries are natural or Indian-made accumulations of shell material and detritus. When of Indian origin, they are referred to as Indian middens (plate 54) in reference to one theory of their origin. *Midden* is a term meaning a dunghill or refuse heap, especially of a primitive habitation. The middens have distinct shapes, often being circular or horse-shoe shaped. Whether of Indian origin or natural, shell mounds have similar flora, fauna, and dynamics.

A key environmental parameter determining the composition of the mounds is the presence of calcium from the shells. Several calcium-loving species, called calciphytes, along with species of the adjacent maritime forests and salt shrub thickets occur. The calciphytes include buckthorn (*Rhamnus caroliniana,* plate 184), southern sugar maple (*Acer saccharum* subsp. *floridanum*), white basswood (*Tilia heterophylla,* plate 181), the rare shell-mound buckthorn (*Sageretia minutiflora,* plate 55), and crested coral-root (*Hexalectris spicata,* plate 180).

There is a floristic similarity between shell mounds and marl forests as both are influenced by the calcium in the soil. For example, buckthorn, southern sugar maple, and white basswood are often present in both gardens.

THE RIVER GARDENS

Brackish Marshes

Brackish marshes (plate 57) generally have a dense growth comprised of a few species with grasses and sedges dominating. Brackish marshes are a transition community between salt and freshwater marshes. It is, however, a distinct community even though it contains species of both the salt and freshwater marshes. The reason for this is that some species can grow in more than one system. Needle rush can live in salt, brackish, or freshwater conditions and is found in all three marshes. Another species, smooth cordgrass, dominates the salt marshes because it can tolerate being flooded twice daily by salt water. It can grow in both brackish and freshwater situations, but does so rarely because it cannot compete with other species in these habitats. Big cordgrass (*Spartina cynosuroides*) can grow in either salt, brackish, or freshwater marshes, but is more common in brackish marshes where it often forms dense stands. On the other hand, wild rice (*Zizania aquatica,* plate 59) and saw-grass (*Cladium jamaicense,* plate 64) can grow in either fresh or brackish conditions. These examples indicate why there is often no clear delineation between the three marshes, especially between brackish and fresh water which have many overlapping species.

In the coastal estuaries, needle rush (*Juncus roemerianus,* plate 60) sometimes forms what is close to a natural "monoculture" upland from the salt marshes. Here, fresh water from rains runs off the higher ground into the marsh, partially diluting the salt water and creating brackish conditions where needle rush thrives. Also, the slight increase in elevation means it is only periodically flooded by salt water, further creating the brackish condition.

Brackish marshes extend up the rivers of the Lowcountry where they grade into the freshwater marshes as the salinity decreases. Along the rivers, however, instead of monoculture stands of needle rush, grasses, and sedges dominate. Many of the conspicuous grasses and rushes of this community are included in this book as wildflowers. Prominent are the sedges—soft-stem bulrush (*Scirpus validus*), leafy three-square (*Scirpus robustus*), and saw-grass (*Cladium jamaicense*)—and the grasses—big cordgrass (*Spartina cynosuroides*), wild rice, and southern wild rice (*Zizaniopsis miliacea,* plate 58).

Tidal Freshwater Marshes

Freshwater marshes occur along the tidal rivers, inland along pond and lake margins, and in managed impoundments. Of the two types of freshwater marshes, the inland and tidal, the latter covers by far the greater area in the Lowcountry.

Tidal freshwater marshes (plate 65) are much more diverse ecologically and floristically than either the salt or brackish marshes. In the freshwater tidal marshes along the Cooper River in Berkeley County, over one hundred species of vascular plants were identified. A similar diversity occurs in the marshes of the other rivers in the Lowcountry. The floristic composition varies from site to site within a river as well as between rivers. Zonation may exist within a site, but is not repeated consistently from site to site.

In the original rivers of the Lowcountry, tidal freshwater marshes occurred as fringes along the rivers where tidal swamp forests bordered. Today, however, the majority of tidal freshwater marshes occur in former rice fields. The rice fields, in turn, were former tidal cypress-gum swamps. The swamps along every tidal river along the Carolina coast, where there was at least a 3-foot difference in tidal amplitude, were ultimately put into rice production. When rice growing ended in the early 1900s, the fields followed two fates: (1) either the banks around the fields were maintained and water control structures were used to manipulate vegetation growth by selecting a water regime that encouraged growth of species that attracted waterfowl (generally for hunting), or (2) the fields were abandoned, allowing nature to take its course. It is these abandoned fields that today support the greatest acreage of tidal freshwater marshes.

As in the brackish marshes, the species that characterize this marsh community are those with their leaf-bearing stems or leaves extended above the water. They include various rushes, sedges, cat-tails, and hydrophytic flowering species. These flowering species, most of which do not occur in the brackish or salt marshes, make this community one of the greatest wildflower gardens of the Lowcountry. Although the flowering species do not dominate the system, they are sufficiently common to add a distinctive beauty and color to the marsh-scape. The best way to view this garden is by boat, but visitors should be aware of the potential conflict with biting insects and snakes.

Some of the more conspicuous wildflowers of this garden are cardinal flower (*Lobelia cardinalis*), spider-lily (*Hymenocallis crassifolia*), alligator-weed (*Alternanthera philoxeroides*), eryngo (*Eryngium aquaticum*), swamp rose (*Rosa palustris*), ground-nut (*Apios americana*), native wisteria (*Wisteria frutescens*), water parsnip (*Sium suave*), water hemlock (*Cicuta maculata*), swamp rose mallow (*Hibiscus moscheutos*), seashore mallow (*Kosteletskya virginica*), arrow arum (*Peltandra virginica*), pickerelweed (*Pontederia cordata*), fragrant ladies' tresses (*Spiranthes odorata*), water-spider orchid (*Habenaria repens*), lizard's tail (*Saururus cernuus*), blue flag iris (*Iris virginica*), jewelweed (*Impatiens capensis*), and climbing hempweed (*Mikania scandens*).

Numerous grasses, sedges, and rushes form the dominant plants of this marsh, most of which also occur in the brackish marsh: wild rice, southern wild rice, big cordgrass, cat-tails (narrow and common), and saw-grass. Two freshwater species are giant beard grass (*Erianthus giganteus*) and marsh bulrush (*Scirpus cyperinus*).

Along the margins of the marshes where they border the open water, species of the freshwater aquatic gardens may occur.

Tidal Freshwater Swamp Forests

Tidal freshwater swamp forests (plate 96) occur from the upper limit of freshwater tidal influence to the brackish water line along the lower reaches of the great brownwater rivers that arise in the mountains (Santee, Great Pee Dee and Savannah) and along the numerous, shorter, blackwater rivers (Ashepoo, Cooper, Combahee, and Waccamaw) that arise in the coastal regions. The best relatively undisturbed examples of this community are along the Waccamaw River in Georgetown and Horry counties and along the Black River in Georgetown County. Small stands occur scattered along the other rivers. Most of the original tidal swamp forests were converted to tidal rice fields beginning in the middle 1700s; today, many of these fields support secondary forests that became established after the fields were abandoned.

The canopy members of the relatively undisturbed tidal swamps are essentially the same as in the bottomland swamp forests discussed later; bald cypress, tupelo gum, swamp-gum, red maple, and water ash are common. Swamp-gum (*Nyssa biflora*) tends to be more common in the tidal swamps than in the bottomland swamps. Virginia willow, dwarf palmetto, *Viburnum* ssp., swamp willow (*Salix caroliniana*), red bay, and water-elm may occur in the subcanopy and shrub layer; the herb layer generally consists of the same species as the tidal marshes.

The wildflowers of the tidal freshwater swamps are essentially the same species that occur in the bottomland swamp forests and tidal freshwater marshes; accordingly, no color plates of species are included as a separate section for this community.

Stream Banks

As the larger, brownwater rivers wind their way through the flat Lowcountry, sand and mud carried by the current settles out in points along the river's edge forming sandbars. On the bars a distinct community forms—the stream banks (plate 97), dominated by shrubs and trees characteristic of floodplains or wet places such as tag alder (*Alnus serrulata,* plate 99), river birch (*Betula nigra,* plate 98), American sycamore (*Platanus occidentalis,* plate 102), swamp cottonwood (*Populus heterophylla*), button-bush (*Cephalanthus occidentalis,* plate 100), water ash (*Fraxinus caroliniana,* plate 101), and swamp willow (*Salix caroliniana*).

This is a temporary community since the trees and shrubs are usually not able to reach large size before the stream changes and the sandbars wash away or become altered by currents or flooding. On the bars that do persist, succession leads to replacement of this community by the adjacent floodplain community.

THE FRESHWATER AQUATIC GARDENS

Freshwater Aquatics

The freshwater aquatic gardens (plate 103) include the true aquatic plants: (1) the submerged and anchored; (2) the floating-leaved, anchored; and (3) the floating (that has no contact with the soil). The aquatics find their greatest development in lakes, ponds, freshwater sounds, canals, abandoned rice fields, and sluggish streams. Much of the habitat today for the aquatics is human-created. Reservoirs such as Lake Moultrie, completed in 1942, harbor along their shores a rich, aquatic growth. Inland swamps that were dammed to create water reservoirs for growing rice now support aquatic populations, and many multipurpose canals dug for whatever reasons provide habitat for the aquatics. Mention has already been made of the abandoned tidal rice fields that support aquatics on the fringe of the marsh vegetation. The aquatics are also found in deepwater pockets in many of the inland swamp forests where the water level is maintained year around by dams and in ox-bow lakes formed along the major rivers.

A number of submerged, anchored aquatics occur in the Lowcountry. Often their flowers project above the water's surface. Two species with conspicuous flowers are fanwort (*Cabomba caroliniana*) and waterweed (*Egeria densa*). The floating aquatics are represented by water hyacinth (*Eichhornia crassipes,* plate 109) and four genera of the duckweed family (*Lemnaceae*): duckmeat (*Spirodela,* plate 104), duckweed (*Lemna*), water-meal (*Wolffia*), and bog-mat (*Wolffiella*).

The most spectacular of the aquatics are the floating-leaved, anchored species with their showy flowers borne above the water's surface. Most prominent of these is fragrant water-lily (*Nymphaea odorata,* plate 111). Other species are cow-lily (*Nuphar luteum,* plate 106), frog's-bit (*Limnobium spongea,* plate 113), water-shield (*Brasenia schreberi,* plate 114), pondweed (*Potamogeton pulcher,* plate 112), floating-heart (*Nymphoides aquatica,* plate 105), and floating bladderwort (*Utricularia inflata* var. *inflata,* plate 108).

One species that does not fit any of the categories above is purple bladderwort (*Utricularia purpurea,* plate 107). It has purple flowers held above the water by a mat of submerged, free-floating stems and leaves.

THE PEATLAND GARDENS

Peatlands are wetlands where soils are peat. Peat is the partially or incompletely decomposed remains of dead plants and, to a lesser extent, animals. Peatlands develop where there is a net gain in organic matter over time; decomposition cannot exceed production if peatlands are to form. The primary condition that slows decomposition is water; in water, dead plants and animals decompose at a much slower rate than when they are exposed to both air and moisture. The difference in the decomposition rate is due to the availability of oxygen. Most bacteria and fungi that decompose organic matter need oxygen for respiration. In saturated conditions, no atmospheric oxygen is available. The accumulation of peat is also helped if the water moves slowly; slowly moving water does not carry away the accumulation of organic matter. Acids that build up as by-products of respiration further inhibit decomposition in bogs because bacteria and fungi do not work as effectively under acid conditions. Bog waters develop a dark color from the acids that are not washed away by moving water.

Pocosins

Pocosins (plate 115) are one of the coastal plain's botanical treasures and the dominate peatland garden in the Lowcountry. Called pocosins by the Algonquin Indians, meaning "swamp-on-a-hill," these are dominated by a dense mix of evergreen shrubs, vines, and scattered trees. In some pocosins, the vegetation is so thick and laced with bamboo-vine (*Smilax laurifolia,* plate 134) that traversing the pocosin is almost impossible. Often one has to go down "on all fours" to get below the dense tangle of bamboo-vine to make any headway. But for those who venture forth, the reward is a garden of pocosin wildflowers that, though less diverse, is equal in beauty to the other gardens.

Pocosins occur on the Atlantic Coastal Plain from Virginia to Georgia. In the outer coastal plain of South Carolina, pocosins develop mainly on flat or depressed sites in the uplands between rivers and streams and in the swales between or parallel to sandhill ridges. Many of these upland depressions are Carolina bays. In these sites, during the rainy season, the soil becomes water-logged for extended periods; during periods of drought the soil becomes fairly dry as the water table falls many feet below the surface. Natural drainage is poor because the pocosins are some distance from large streams.

The vegetation of pocosins is a dense growth of shrubs associated with scattered trees. Diversity is not great since few species can adapt to the mineral-poor, acid soils and long hydroperiods. The dominant tree is pond pine (*Pinus serotina,* plate 116); loblolly bay (*Gordonia lasianthus,* plate 133), red bay (*Persea borbonia*), and sweet bay (*Magnolia virginica,* plate 127) occur as associates. In the Lowcountry, the most frequent shrubs are the evergreens fetterbush (*Lyonia lucida*), inkberry (*Ilex glabra*), sweet gallberry (*Ilex coriacea*), and the deciduous shrubs titi (*Cyrilla racemiflora,* plate 131) and honeycup (*Zenobia pulverulenta,* plate 126), all growing with bamboo-vine. Because of the evergreen shrubs, the pocosins are called by some workers "evergreen shrub bogs." The shrubs in some sites are short (2–3 feet) with scattered pond pine and are called low pocosins; in other sites, both shrubs and trees are taller and the trees denser. These are the high pocosins. In some pocosins, tall zones exist around the margin with the low pocosin growth in the center. Poison sumac (*Rhus vernix,* plate 128) often grows along the margins of pocosins. Two rare evergreen, ericaceous shrubs—leather-leaf (*Cassandra calyculata,* plate 118) and lamb-kill (*Kalmia angustifolia* var. *caroliniana,* plate 124)—are found along the edges of pocosins.

Fires increase the habitat diversity of pocosins by unevenly burning the peat and creating depressions below the water table.

These depressions fill with sphagnum and herbaceous species such as yellow trumpet pitcher-plant (*Sarracenia flava,* plates 372 and 373), frog's britches (*S. purpurea,* plate 123), and white arum (*Peltandra sagittaefolia,* plate 132).

For thousands of years, peat has been building up in these pocosins to a depth of 10 feet. During droughts peat is susceptible to fires, and most pocosins burn about every ten to thirty years. Pocosin plants have adapted to this fire cycle by sprouting vigorously from their rootstocks that are afforded protection from fire by being buried in the deeper, wet zone of the peat. Pocosins are also mineral-poor since they receive water mostly from rainfall or drainage through coarse sands.

Occasionally associated with pocosins, as a special microhabitat, are the seepage bogs. Where the adjacent pine flatwoods slope toward the pocosins (or a swamp forest), a permanent, wet zone develops. Rainwater that percolates into the sandy flatwoods soil hits a confining layer that forces the water to move laterally until it emerges as seepage at the base of the slope. The soil is permanently saturated (except in extreme droughts) and very springy. A rich variety of bog species occurs in the seepage bogs, many the same ones that occur in the longleaf pine savannas. Seepage bogs can be distinguished from the pine savannas by the presence of the pitcher-plants. Hooded pitcher-plant (*Sarracenia minor*) is the dominant species in the pine savannas, but is absent or rare in the seepage bogs. In seepage bogs, sweet pitcher-plant (*S. rubra,* plate 122) and yellow trumpet (*S. flava*) are the dominant species. Sweet pitcher-plant does occur in the savannas, but grows poorly. When growing in seepage bogs, however, it is robust. Fire is also important in maintaining this microhabitat; lack of fire allows adjacent pocosin shrubs to dominate the site.

Pocosins today are being studied intensely as natural reservoirs for water and as habitats for plants and animals. During times of drought the lower layers of peat hold water that is available to animals or that can be slowly released into the surrounding areas. The dense vegetation also provides habitats for animals that avoid people, such as the endangered black bear and pine barrens treefrog. Pocosins have had a long history of human use; timbering, drainage for agricultural use and timber plantations, peat mining, and urban development all have greatly reduced pocosin habitat. Fortunately, steps are being taken today by concerned organizations to preserve the pocosins.

THE BOTTOMLAND FOREST GARDENS

Bottomland forests occupy the flat, alluvial floodplains above the upper limit of tidal influence that flanks the river systems which traverse the Lowcountry. These forests and their associated fauna comprise remarkably productive riverine communities adapted to fluctuating water levels. In the face of intensive land-use of the adjacent uplands, the bottomland forests today serve as refuges for floodplain species as well as upland species of wildlife.

Two major types of floodplain rivers traverse the Lowcountry: brownwater and blackwater. Brownwater rivers originate in the mountain and piedmont areas and have broad floodplains. The brown color of the water comes from silt and clay suspended in the water as a result of erosion from the piedmont and mountains. Most of these brownwater rivers, such as the Santee, have periods of sustained high flow resulting from the cumulative effects of many tributaries and distant rainfall.

Blackwater rivers and tributary streams originate in the coastal plain and receive most of their water from local rain. They have narrower, less well-developed floodplains. Unlike the brownwater rivers, the blackwater rivers may have dry periods during which discharge may be very low. The term blackwater comes from the relatively clear, but highly colored water due to the presence of organic acids derived from decaying leaves. Examples of blackwater rivers are the Cooper, Ashley, Combahee, Ashepoo, New, Four Holes, Waccamaw, and Black.

Two major bottomland communities occur on the floodplains of the brownwater and blackwater rivers above the zone of tidal influence: bald cypress–tupelo gum swamp forests and bottomland hardwoods.

Bald Cypress–Tupelo Gum Swamp Forests

The bottomland bald cypress–tupelo gum swamps (plate 135) represent the forested community least disturbed in the Lowcountry. Still, only a few original growth stands remain.

In swamps where the land is flooded almost continuously, bald cypress (*Taxodium distichum,* plates 135 and 137) and tupelo gum (*Nyssa aquatica,* plate 160) may exist together, or each may occur separately in pure stands. Pure tupelo stands, however, often follow the clear-cutting of cypress-tupelo stands. Knee formations of cypress, reaching 6 feet or more, and buttress and knee formations of cypress and tupelo gum are more pronounced in deep sloughs. Shrubs and herbs are sparse because of the flooded conditions and dense canopy. A distinctive swamp microhabitat is the herbs growing on floating logs and stumps. Species characteristic of this microhabitat are skullcap (*Scutellaria lateriflora,* plate 157), St. John's-wort (*Hypericum walterii*), false nettle (*Boehmeria cylindrica,* plate 156), and clearweed (*Pilea pumila*).

The epiphytes green-fly orchid, Spanish moss, and resurrection fern grow on branches of trees. Vines such as cross vine (*Anisostichus capreolata*), coral greenbrier (*Smilax walterii*), supplejack (*Berchemia scandens*), and poison ivy (*Rhus radicans*) exhibit pronounced growth, especially at the margin of the swamps. Ladies'-eardrops (*Brunnichia cirrhosa,* plate 151), a rare woody vine, can be found on the margin of the swamp forests near lakes or ponds. Often the swamps have lakes within them (remnants of old streams) where members of the freshwater aquatics and freshwater marshes occur.

As the depth and duration of flooding decreases, more mesic trees such as red maple, water ash, swamp-gum, and cottonwood form a subcanopy. As the wet soil becomes more exposed, shrubs become common, including Virginia willow (*Itea virginica*), swamp dogwood (*Cornus stricta*), leucothoe (*Leucothoe racemosa*), and possum-haw (*Ilex decidua*). Southern rein-orchid (*Habenaria flava*) occurs on the edges of muddy sloughs. Where favorable conditions exist, lizard's tail (*Saururus cernuus*), blue flag iris (*Iris virginica*), butterweed (*Senecio glabellus*), aquatic milkweed (*Asclepias perennis*), water willow (*Justicia ovata*), and golden-club (*Orontium aquaticum*) flourish, all adding color to the swamp.

A great diversity of species occurs from swamp to swamp. Age of the forest, past timbering, degree of flooding, soil composition, and freedom from disturbance have all contributed to the composition of today's swamps. Only Beidler Forest in Four Holes Swamp harbors a significant stand of original growth forest; the remaining swamp forests of the Lowcountry are primarily secondary growth.

Hardwood Bottoms

Hardwood bottoms (plate 162) occur on floodplains somewhat elevated above adjoining cypress-gum swamps. Although flooded for a considerable period, the surface through much of the year is dry. Hardwood bottoms are extremely diverse floristically and quite variable from one site to another. The vegetation of the hardwood bottoms is usually very dense, and in the more undisturbed sites, trees reach large sizes, many over 3 feet in diameter. Small trees and shrubs are frequent and woody vines luxuriant. In the drier sites a rich, herbaceous flora flourishes. In areas that have been logged, trees are much smaller, but still dense. Often hardwood bottoms are a narrow strip of land between the adjacent uplands on one side and the swamp forests on the other; at other times they may be a broad expanse of land between the same two communities.

Many trees characterize this community: sweet-gum (*Liquidambar styraciflua,* plate 173), loblolly pine (*Pinus taeda*), overcup oak (*Quercus lyrata*), water oak (*Q. nigra*), willow oak (*Q. phellos*), swamp chestnut oak (*Q. michauxii*), laurel oak (*Q. laurifolia*), cherry-bark oak (Q. *falcata* var. *pagodaefolia*), ash (*Fraxinus* ssp.), American sycamore (*Platanus occidentalis,* plate 102), American holly (*Ilex opaca*), American elm (*Ulmus americana*), and hackberry (*Celtis laevigata*), among others.

A subcanopy of young canopy species, including ironwood (*Carpinus caroliniana*), can be found. Numerous shrubs characterize the community: swamp dogwood (*Cornus stricta,* plate 145), arrowwood (*Viburnum dentatum,* plate 167), and elderberry (*Sambucus canadensis,* plate 155)—species that also occur in the adjacent swamp forests.

Woody vines are especially prominent and include cow-itch

(*Campsis radicans,* plate 171), poison ivy (*Rhus radicans,* plate 159), supplejack (*Berchemia scandens,* plate 142), climbing hydrangea (*Decumaria barbara,* plate 170), and muscadine (*Vitis rotundifolia*).

Grasses, rushes, sedges, and typical wildflowers form a rich herbaceous layer in the drier sites. Grasses and sedges often form a dense ground layer and can be used to separate the hardwood bottoms from the adjacent swamp forests. Many of the wildflowers from the swamp forests also grace the hardwood bottoms, but several additional species can be found, including Easter lily (*Zephyranthes atamasco,* plate 165) and Jack-in-the-pulpit (*Arisaema triphyllum,* plate 163).

An interesting microhabitat in the hardwood bottoms is the "windthrow" community. The bottomland trees have shallow but broad root systems; when blown down, the uplifted soil clings to the roots, which is now in a sunlit area because of the broken canopy. This provides an entry for numerous herbs. Often two weedy species from the uplands find a home here: pokeweed (*Phytolacca americana,* plate 441) and fireweed (*Erechtites hieracifolia*).

THE DECIDUOUS FOREST GARDENS

The eastern deciduous forest lies in the heart of eastern North America: from the Blue Ridge and southern Appalachian foothills in the Carolinas and Virginia across the Cumberland Mountains of Kentucky and Tennessee, to the ridges and valleys from New York and Pennsylvania southwestward, across much of Ohio and Indiana and into Illinois. The dominant characteristic of the deciduous forests is division of the year into distinct seasons. Of critical importance is the continuous below-freezing winter temperature in the north and occasional frost in the south. Under these conditions evolved the deciduous habit: the shedding of broad leaves of angiosperms during the winter. Retaining broad-shaped leaves in the winter could result in dehydration because the broad leaves lose more water than is available in the frozen winter soil.

In the Carolina Lowcountry, deciduous forests are not a dominant feature of the landscape. Most deciduous forests that occurred on upland sites were cleared for agriculture in the 1700s and 1800s. When the fields were abandoned, pine or pine–mixed hardwoods replaced them. In time, however, the pine–mixed hardwoods will be replaced by the climax hardwoods. Today good stands of deciduous forests exist along the ravines of creeks and rivers and on some upland sites, that, for whatever reasons, escaped being disturbed.

Wildflower enthusiasts love the deciduous forests, especially the beech forests, because of the abundance of spring wildflowers. Virtually all herbs in undisturbed deciduous forests are perennials with underground storage organs. They are able to quickly produce flowers and seeds in early spring before the canopy develops. During this time they also carry on photosynthesis and store food for the next year. Once the canopy closes during the summer, photosynthesis essentially ceases and aerial parts die. But in this short time, these beautiful flowers (trilliums, bloodroot, green-and-gold, and wild geranium) grace the woodlands. In addition, there are four different types of deciduous gardens that occur in the Lowcountry.

Marl Forests

Marl forests (plate 174) occur on mesic sites over shallowly buried or exposed marl formations. A rare community in the Lowcountry, its few remaining sites are in danger of being disturbed. It contains a unique and diverse assemblage of trees, shrubs, and herbs, many of which are thought to be calciphytes. The calcium in the soil comes from the calcium carbonate of sea shells that comprise the marl formation laid down as a marine deposit. Some of the species occur as common members of other deciduous communities; some, however, appear to be confined to the high-calcium soils. Calciphytes, such as blackstem spleenwort (*Asplenium resiliens,* plate 175) and alumroot (*Heuchera americana,* plate 177), only grow where marl occurs, as along Wadboo Creek in Berkeley County.

Trees that characterize the marl forests include yellow chestnut oak (*Quercus muehlenbergii*), hop hornbeam (*Ostrya virginiana,* plate 183), white basswood (*Tilia heterophylla,* plate 181), slippery elm (*Ulmus rubra*), buckthorn (*Rhamnus caroliniana,* plate 184), and southern sugar maple (*Acer saccharum* subsp. *floridanum*). Elements of the other deciduous forests include redbud, dogwood, and red cedar. Herbaceous species are common and include many found in other deciduous communities. Several species, besides those mentioned earlier, that seem to be more common in this community include crested coral-root (*Hexalectris spicata,* plate 180), shadow-witch (*Ponthieva racemosa*), *Elytraria caroliniensis* (plate 179), and thimbleweed (*Anemone virginiana,* plate 178).

The rarity of this community warrants research into its ecology and preservation of remaining sites.

Beech Forests

Beech forests (plate 185) in the Lowcountry occur on moist, well-drained soils found on north-facing river bluffs, seepage slopes, and ravine slopes. Some of its species are southern in origin, including a few bottomland hardwood species mixed in with the typical mesophytic ones. Others migrated from the older land area of the interior. Although it is not a common community in terms of acreage, numerous good sites occur scattered throughout the Lowcountry.

The best indicator species is American beech (*Fagus grandifolia,* plate 208); its associates differ from site to site and include the following canopy trees, many of which occur in the oak-hickory gardens: tulip tree (*Liriodendron tulipifera,* plate 200), white oak (*Quercus alba*), swamp chestnut oak (*Q. michauxii*), sweet-gum (*Liquidambar styraciflua,* plate 173), holly (*Ilex opaca*), bull bay (*Magnolia grandiflora,* plate 37), spruce pine (*Pinus glabra*), loblolly pine (*Pinus taeda*), mockernut hickory (*Carya tomentosa,* plate 242), and red maple (*Acer rubrum,* plate 139). Occasionally umbrella tree (*Magnolia tripetala*), a piedmont species, is found. Understory species include dogwood, witch-hazel (*Hamamelis virginiana,* plate 210), redbud, horse sugar (*Symplocos tinctoria*), ironwood (*Carpinus caroliniana*), and sweet shrub (*Calycanthus floridus,* plate 198).

But it is the great diversity of herbs, which shows a tendency toward a piedmont and mountain distribution (especially the spring ephemerals), that makes the beech gardens a wildflower paradise. In early spring, before the leaves have appeared on the deciduous trees, the forest floor is a beautiful garden of herbaceous wildflowers. Some noted spring ephemerals include bloodroot (*Sanguinaria canadensis,* plate 186), may-apple (*Podophyllum peltatum,* plate 189), green-and-gold (*Chrysogonum virginianum,* plate 192), bellwort (*Uvularia perfoliata,* plate 199), little sweet betsy (*Trillium cuneatum,* plate 190), wild geranium (*Geranium maculatum,* plate 202), and wild ginger (*Asarum canadense,* plate 197). A rich fern growth also graces this garden. The herbaceous parasites beech-drops (*Epifagus virginiana,* plate 209) and cancer-root (*Conopholis americana,* plate 193) are also found in the beech forest. The dense canopy precludes a dense shrub or herbaceous layer from developing, making walking easy throughout the year.

Oak-Hickory Forests

Oak-hickory forests (plate 211) occur on upland slopes between rivers and tributaries. These sites are drier than the beech sites, resulting in dominance by the more drought-resistant oaks and hickories. This is a complex community, abundantly distributed and difficult to characterize, with much variation among sites. Undisturbed examples are hard to locate because of intense forestry and other disturbances. It is also difficult to separate this community from adjacent forest types because of the overlap of species. Along the coast, it grades into the maritime forests, while in the uplands it grades into the beech forests as moisture increases, and into the xeric communities as moisture decreases. Variations in soil, moisture, incidence of fire, past and present disturbance, and exposure all create environmental conditions favorable for many kinds of trees that vary from site to site; however, the oaks and hickories are the codominants. The poorer sites are usually typified by mockernut hickory (*Carya tomentosa,* plate 242), blackjack oak (*Quercus marilandica*), and post oak (*Q. stellata*); the better sites by white oak (*Q. alba*), cherry-bark oak (*Q. falcata* var. *pagodaefolia*), water oak (*Q. nigra*), and swamp chestnut oak (*Q. michauxii*). This selectivity has prompted some workers to subdivide the oak-hickory forests into the "dry" and "mesic" subtypes. A high predominance of sweet-gum and pines indicates past disturbances. Other canopy trees that occur are loblolly pine, shortleaf pine (*P. echinata*), black gum, red maple, and tulip tree.

Many showy subcanopy trees occur as components of the oak-hickory forests: flowering dogwood (*Cornus florida,* plate 212), sassafras (*Sassafras albidium,* plate 213), horse sugar (*Symplocos tinctoria,* plate 214), redbud (*Cercis canadensis,* plate 215), fringe-tree (*Chionanthus virginicus,* plate 221), crab-apple (*Malus angustifolia,* plate 223), and bigleaf snowbell (*Styrax grandifolia,* plate 222).

The herbaceous growth is so varied from site to site that listing typical species is difficult. In the drier sites, legumes and grasses may dominate, while in others more showy composites may dominate, especially in the fall. Nonetheless, numerous showy wildflowers grace the oak-hickory forests, including dwarf iris (*Iris verna,* plate 217), lousewort (*Pedicularis canadensis,* plate 220), white milkweed (*Asclepias variegata,* plate 226), Indian pink (*Spigelia marilandica,* plate 227), ruellia (*Ruellia caroliniensis,* plate 231), butterfly-pea (*Clitoria mariana,* plate 233), large false foxglove (*Aureolaria flava,* plate 238), and coral honeysuckle (*Lonicera sempervirens,* plate 218). In sites with dense shrubs and young trees, herbs may be sparse.

This community has few wildflowers unless a mature, relatively undisturbed example can be located. Most oak-hickory sites in the Lowcountry have been disturbed, resulting in a dense subcanopy and shrub layer that makes ingress difficult, especially during the summer and fall when vegetative growth is advanced.

Sandy, Dry, Open Woods

The sandy, dry, open woods (plate 243) community is not recognized in standard texts as a distinct community. Indeed, it is difficult to describe its major components; nonetheless, there is no difficulty recognizing the community in the field and, with sufficient study, it could be given formal status. It is probably a successional community that develops on sand ridges following agricultural abandonment, or from disturbance in dry oak-hickory forests or dry pinelands.

Canopy trees are scattered and consist of “dry” oaks and hickories with occasional loblolly pines; shrubs are sparse. The herb layer is open and best defines the community. Invariably several of the following herbaceous wildflowers occur in the community: blue curls (*Trichostema dichotomum,* plate 260), southern gaura (*Gaura angustifolia*), cottonweed (*Froelichia floridana,* plate 255), blazing-star (*Liatris elegans,* plate 261), horse mint (*Monarda punctata,* plate 257), blue star (*Amsonia ciliata,* plate 244), three species of lupine (*Lupinus perennis, L. villosus,* and *L. diffusus*), wild pink (*Silene caroliniana,* plate 248), prickly-pear (*Opuntia compressa,* plate 259), piriqueta (*Piriqueta caroliniana,* plate 252), and queen’s-delight (*Stillingia sylvatica,* plate 250). Several of these herbs also occur in the xeric sandhills.

THE PINELAND GARDENS

Imagine, if you will, the hundreds of thousands of acres of almost pure stands of longleaf pine that greeted the early settlers as they moved inland from the coast of the Atlantic and Gulf coast states. Here were unbroken longleaf pine forests with trees over 120 feet tall and 40 or more inches in diameter. This aristocrat of southern pines was more abundant than shortleaf, loblolly, and slash combined, comprising 90 percent of the original pine forests. With their tall, straight, and slender trunks, the original trees must have impressed the early settlers. But these giants are now history. The longleaf came to have more uses than any other tree in North America. The original forest fell to the ax of the lumberjack, but so vast were the forests that no one thought of replanting; without replanting, the longleaf forests were doomed. Several reasons are cited for the lack of sufficient natural restoration. There was a decrease in the natural fires that provide open, mineral soil necessary for good seed germination. In the early 1900s, longleaf saplings were popular as Christmas trees, further decreasing the chance for restoration. Abandoned fields that were cut out of the original stands naturally reseeded with loblolly pine, a more prolific seeder. Again, a decrease in natural fires helped to maintain the loblolly forests. Another major problem with longleaf restoration was the feral hogs that ranged over much of the pine country. A single hog with a taste for pine roots could destroy hundreds of seedlings in a day. Also, the original method of obtaining crude turpentine from longleaf pines was "boxing," which frequently killed the tree in about ten years.

Even with conservation methods applied to modern forestry, longleaf pine occupies less than a tenth of its original area. Its main stronghold is in national forests where management policy is to replant longleaf in any site where it is timbered. Small holdings of longleaf exist in sanctuaries such as the Webb Wildlife Center and on large hunting preserves; here one can imagine what the original longleaf forests were like. Below is a description of the major communities which are dominated by longleaf pine.

Xeric Sandhills

The xeric sandhills (plate 262), characteristic of the deep, coarse sandhills, may be recognized by the presence of two trees: longleaf pine (*Pinus palustris,* plate 263) and turkey oak (*Quercus laevis,* plate 277). The presence of other trees indicates some condition has changed, resulting in an increase in soil moisture. In the original pine forests of the sandhills, frequent fires maintained the subclimax pine forests by not allowing the accumulation of organic matter in the soil which is essential to the establishment of hardwoods. Today it is probably the absence of a litter source (no large pines), more than fire, that keeps organic matter from building in the soil. The Lowcountry xeric sandhills (often called "deserts in the rain") are today vegetated by scattered, small longleaf pines and turkey oak, or on some sites with only turkey oak.

Xeric sandhills are not common in the Lowcountry. Good examples occur in Horry County on the sand ridges that border the Carolina bays in Lewis Ocean Bay Heritage Preserve and in isolated tracts along some of the larger rivers (for example, the Savannah, Little Pee Dee, and Waccamaw). Other isolated sites occur in the Honey Hill section of the Francis Marion National Forest in Berkeley County, near a small community known as Sandy Island in Orangeburg County, and on the Tillman Sand Ridge Heritage Preserve in Jasper County.

Very few shrubs and herbs are able to survive in this habitat. The coarse, sandy soil has little water-holding capacity, and soon after a rain the water moves down through the soil—out of reach for most plants. It also carries with it what few minerals are present. Under these conditions it is surprising that any plants can survive, but some do, adding a touch of color to the background of bare, white soil.

Two characteristic shrubs are dwarf huckleberry (*Gaylussacia dumosa,* plate 266) and rosemary (*Ceratiola ericoides,* plate 281). Herbs include sandhills milkweed (*Asclepias humistrata,* plate 271), sandhills baptisia (*Baptisia cinerea,* plate 270), tread-softly (*Cnidoscolus stimulosus,* plate 267), thistle (*Carduus repandus,* plate 273), gerardia (*Agalinis setacea,* plate 279), and wire plant (*Stipulicida setacea*).

In southern Jasper County at the Tillman Sand Ridge Heritage Preserve, a variant of the xeric sandhills occurs that contains some xeric species typical of more southerly locations and unique to this site in the state. The site borders the Savannah River, and the sands are fluvial in nature. Rare species include rose dicerandra (*Dicerandra odoratissima,* plate 280), warea (*Warea cuneifolia,* plate 274), soft-haired coneflower (*Rudbeckia mollis,* plate 275), and gopher apple (*Chrysobalanus oblongifolius,* plate 269), a low shrub.

Longleaf Pine Flatwoods

The typical pine flatwoods (plate 282) are dominated by a canopy of tall, longleaf pines. The terrain is essentially flat to very gently rolling with a sandy soil and high water table. Although longleaf pine characterizes the community, loblolly and slash pine may occur.

Fire is frequent, but not as frequent as in the longleaf pine savannas (see below); accordingly, a well-developed shrub layer and understory may develop. Under higher fire frequency, the shrubs and understory species are kept in check. Because of the variation in the understory and shrub layers from site to site, the pine flatwoods are difficult to characterize. More common understory species include sweet-gum, blackjack oak (*Quercus marilandica*), and black gum (*Nyssa sylvatica*); shrubs include sweet bay, wax myrtle, inkberry, and sweet pepperbush (*Clethra alnifolia,* plate 297).

The herbs of the frequently burned pine flatwoods include a variety of grasses, heaths, legumes, and composites, but few of the showy species of the savannas. Grasses include broomstraws (*Andropogon* ssp.); heaths include species of *Vaccinium* and *Lyonia;* legumes include pencil flower (*Stylosanthes biflora*), beggar's lice (*Desmodium* ssp.), *Lespedeza* ssp., lead plant (*Amorpha herbacea,* plate 298), and goat's rue (*Tephrosia virginica,* plate 293); composites include black-root (*Pterocaulon pycnostachyum,* plate 292), asters (*Aster squarrosus, A. tortifolius, A. reticulatus, A. linariifolius,* and *A. concolor*), and goldenrods (*Solidago* ssp.). The ubiquitous bracken fern (*Pteridium aquilinum*) often forms a dense cover in the spring after fire has removed the previous year's debris. The rare American chaff-seed (*Schwalbea americana,* plate 291) occurs in openings of the herb layer. Conspicuous is a low-growing oak known as the running oak (*Q. pumila*).

The pine flatwoods grade into the pine savannas, and distinguishing between the two communities is often difficult. Small, slight depressions with a high clay-pan occur within the pine flatwoods and harbor the typical, showy species that characterize the savannas. At other sites, savannas cover extensive areas and are readily distinguishable from the adjacent pine flatwoods. Two species may be used to distinguish between the two habitats: hooded pitcher-plant (*Sarracenia minor,* plate 324) and toothache grass (*Ctenium aromaticum,* plate 349). Both species require a little more open and moist conditions than normally found in the pine flatwoods and their presence indicates pine savannas rather than flatwoods. Pocosins, cypress savannas, and upland swamps also occur scattered within pine flatwoods.

The pine flatwoods are a fire subclimax community; prolonged absence of fire will lead to the climax community of the area.

Longleaf Pine Savannas

No natural garden equals the pine savannas (plates 316 and 317) for diversity of showy wildflowers. From early spring through late fall, a progression of herbaceous wildflowers graces the Lowcountry with a mixture of colors. Orchids, carnivorous plants, lilies, composites, plus many other groups all find a home in the sunny, pine savannas.

But it is a paradox in that fire, considered such an anathema by many, is responsible for creation and maintenance of the savannas. In the eons before Native Americans settled in the Lowcountry, natural fires started by lightning swept through the pinelands, mostly during the July–August thunderstorm season. Trees and shrubs with their growing tips at fire level were killed. Herbaceous species, with their stems (rhizomes) under ground, were protected. Shortly after the fire, these herbaceous species put up new growth, and what appeared to be a scene of utter desolation quickly became a wildflower garden again. The savannas, protected from fire, quickly succeed to a shrub community, then to a tree-dominated forest. Under a forest canopy, the savanna herbs, which require high light intensity, cannot survive.

Most savannas today occur in national forests or on large plantations as a result of prescribed burning. Small, privately owned savannas are threatened since natural wildfires are quickly put out or do not spread because of barriers such as roads. Also, savannas are threatened by direct utilization for homes or commercial properties, and by drainage canals. These canals lower the water table, allowing invasion of less moisture-tolerant species. Recent data suggests that more than 95 percent of the former acreage of longleaf savannas has been lost. By the next century there may be little significant acreage of this wildflower garden.

One tree that is able to survive the frequent burning of the savannas is sweet bay (*Magnolia virginiana*). It does so by partially burying its stem in the soil; after a fire, it puts up a cluster of new shoots, giving the appearance of shrubs. But if the root-stem system is dug up, one will find a single, enlarged rootstock, testimony to the fact that the "tree" may be many years old even though its above ground stems represent only one or two years growth.

Two types of longleaf pine savannas can be recognized—one (plate 316) dominated by toothache grass and the other (plate 317) by wiregrass (*Aristida stricta*). Both types develop where a fairly level topography and nondraining subsoil (because of an underlying, impeding clay pan) causes a high (or perched) water table in the rainy season. During droughts the soil above the clay pan may become excessively dry. Loblolly pine, slash pine, and pond pine may occur with the longleaf. Only wiregrass and the bog species can tolerate such extreme changes in soil moisture. Wiregrass, however, favors the coarser soils with some slope to allow lateral drainage, resulting in slightly drier conditions; toothache grass and the showy herb species prefer wetter conditions.

It is in the toothache grass savannas that the greatest display of showy herbs occurs. This is the most diverse herb community in the Southeast. Plates of forty-eight species of wildflowers that characterize the toothache grass savannas are included in the color plates.

Pine–Saw Palmetto Flatwoods

The pine–saw palmetto flatwoods (plate 367) are a variant of the longleaf pine flatwoods in which saw palmetto (*Serenoa repens*) dominates the shrub layer. In South Carolina, this community is found only in Jasper and Beaufort counties, the northern limit of saw palmetto; it is more extensive in Florida and Georgia. The canopy consists of longleaf pine on the ridges and slash pine (*P. elliottii*) and/or pond pine in depressions. A subcanopy of oaks is absent or sparse. The shrubs, besides saw palmetto, include typical pocosin and pine flatwood species: sweet bay, red bay, inkberry, sweet gallberry, fetterbush, sweet pepperbush, highbush blueberry, honeycup, stagger-bush (*Lyonia ferruginea*), hairy wicky (*Kalmia hirsuta,* plate 368), and *Vaccinium myrsinites.* The sparse herbaceous layer is a mixture of pine flatwood species.

Fires maintain this community; absence of fires lead to more dominance of shrubs, and periodic fires promote herbs and saw palmettos.

Pine–Mixed Hardwood Forests

The pine–mixed hardwood forests (plate 369) exemplify the history of disturbance in the Lowcountry. This community is successional in nature and is composed of a canopy of various pines with oaks, hickories, sweet-gum, and red maple forming a subcanopy. Or, depending on the successional state of the community, these hardwoods may form part of the canopy. The shrub layer is very thick, dominated by inkberry, wax myrtle, sweet pepperbush, *Lyonia lucida,* and a variety of blueberries and huckleberries. The herb layer (1) may be sparse (because of the thick shrub layer), (2) may contain no unusual herb species, (3) usually contains bracken fern, and (4) may contain a wide variety of submesic to subxeric herbs.

This community is most common on abandoned fields where loblolly pine is the first tree to become established. Loblolly does best in intense sunlight, and if a seed source is nearby, its winged seeds are readily dispersed by wind. In about twenty to twenty-five years the canopy is too dense for pine seedlings to survive, and further pine reproduction is prevented. Hardwood seeds, brought in by wind or animals, can germinate and the seedlings develop in the shaded forest floor. For the next fifty years of so, the pines and hardwoods coexist; ultimately the pines die of old age and the hardwoods become the climax community.

The pine–mixed hardwood forests occur over a wide variety of soil types and can quickly develop following disturbance. It is perhaps the largest in acreage in the Lowcountry; however, it is generally so thick that ingress is difficult and most of the time few wildflowers are present. The trees, shrubs, herbs, and vines that occur in the pine–mixed hardwood forests also occur in other communities.

THE POND CYPRESS GARDENS

Pond Cypress Savannas

Pond cypress savannas (plate 370) occupy flat, poorly drained lands within the longleaf pine forests. They have a slightly longer hydroperiod than the associated pine flatwoods and pine savannas and are dominated by pond cypress, which can tolerate a longer hydroperiod than pines. The cypress canopy is open; red maple and swamp-gum may also be present.

Pond cypress savannas occur scattered throughout the Lowcountry with many of the best examples occurring within Carolina bays. Draining and ditching, along with absence of fire, have reduced the number of pond cypress savannas. Numerous excellent sites still occur in the Francis Marion National Forest.

Few shrubs occur except for some woody hypericums (*Hypericum fasciculatum,* plate 377), myrtle dahoon (*Ilex myrtifolia,* plate 387), and button-bush. The herbaceous flora, blooming through the spring, summer, and fall, is rich with some of the showy species of the adjacent pine savannas, including many of the carnivorous plants. Certain herbs, however, appear to be more common or confined to the pond cypress savannas. These include sneezeweed (*Helenium pinnatifidum,* plate 374), bay blue-flag iris (*Iris tridentata,* plate 375), tall milkwort (*Polygala cymosa,* plate 376), tickseed (*Coreopsis gladiata,* plate 381), yellow trumpet pitcher-plant (*Sarracenia flava,* plates 372 and 373), awned meadow-beauty (*Rhexia aristosa,* plate 379), *Agalinis linifolia, Lobelia boykinii,* and the federally endangered Canby's dropwort (*Oxypolis canbyi*).

Pond Cypress–Swamp-gum Swamp Forests

Pond cypress–swamp-gum swamp forests are dominated by pond cypress or pond cypress and swamp-gum (*Nyssa biflora*) with pond pine (*Pinus serotina*) often an associate; these swamps occur in depressions in the uplands with some water on the surface at least three months a year. The depressions may be limestone sinks, irregular depressions, or Carolina bays.

Good examples of these swamps occur in limestone sinks (plate 382). The herbaceous flora of this system is sparse compared to the riverine and other upland systems; however, numerous shrubs occur along the margins of the depressions. In the Francis Marion National Forest, several populations of pond berry (*Lindera melissifolia,* plates 383 and 384), an endangered species, have been located in sink holes in the Honey Hill and Cainhoy areas. Another member of the laurel family, pond spice (*Litsea aestivalis*), rare throughout its range, is common in these same sinks and elsewhere in the Francis Marion National Forest. Other shrubs in this upland swamp forest include titi, button-bush, dahoon holly (*Ilex cassine,* plate 386), and myrtle dahoon (*Ilex myrtifolia,* plate 387). The rare climbing fetterbush (*Pieris phillyreifolia,* plate 385) is known in South Carolina only from these pond cypress–swamp-gum gardens in the Francis Marion National Forest. The open water is also habitat for the freshwater aquatics with floating bladderwort especially common.

THE RUDERAL GARDENS

The ruderal gardens (plate 388) have become home for countless numbers of native and naturalized species. Ruderal gardens include lawns, roadsides, sidewalks, vacant lots, abandoned fields, garbage dumps, landfills, and railroads where little is left of the natural vegetation. Nonnative species dominate the vegetation, associating with the more-hardy native species that have been able to make the transition to this harsh environment. But many natural environments also harbor these aliens, especially in openings created in the canopy. Popcorn tree (*Sapium sebiferum,* plate 429) seems at home in maritime forests where it is able to compete equally with the native species. But this is the exception as most of the introduced species find it difficult to compete with the native species adapted to their own habitat.

The reader can quickly comprehend the magnitude of this "invasion" by the following: forty-one of the species included in the ruderal gardens are nonnative. Numerous other introduced species occur, but either do not have showy flowers or are too small to photograph well and are not included in this book.

Four features of the ruderal gardens make them worthy of inclusion in this book: (1) they are readily available (one only has to go as far as the backyard or a vacant lot to discover this world of native and naturalized species); (2) the naturalized species are, in many cases, as attractive as the native species (who cannot be impressed by the beauty of common dandelion, wild carrot, woolly mullein, or crimson clover); (3) the ruderal gardens present an everchanging tableau of beauty (the early spring henbit and chickweed give way to the summer prickly sow-thistle, a situation repeated in every habitat these aliens come to dominate); and (4) many of the naturalized species have become part of the lore of the Lowcountry (for instance, common dandelion flowers make wine or branches of Chinaberry are used as a fleabane).

Today, countless acres of the ruderal gardens lace the Lowcountry, adding a patchwork of beauty to compliment the natural gardens. These ruderal gardens will continue to grow at the expense of the native gardens. This is inevitable as urban development spreads in the Lowcountry, especially along the coastal area so valued for resort developments.

Natural History of Selected Groups of Flowering Plants

CARNIVOROUS PLANTS

The carnivorous plants stand alone in the plant world because of their unique method of supplementary nutrition and the manner in which their leaves have become modified through evolution to trap insects and small animals. Plants that adopted carnivory inhabit acidic, mineral-poor soils or waters. The mineral-poor nature of the soil is due in part to the inherent paucity of minerals and in part to the fact that in highly acid soils many minerals are strongly bound to soil particles and are unavailable to plants. In habitats like these evolved the capacity of carnivory which functions as a supplementary means to obtain minerals. From the trapped animals, mostly small insects, digestion of animal protein results in the release of minerals that are absorbed by the plants. As a result of selective pressure, plants that evolved carnivory were able to inhabit the mineral-deficient sites where they found an area less competitive than the surrounding environs.

The carnivorous plants do not obtain food from their prey, only minerals. Photosynthesis occurs as in all green plants. To convert the simple carbohydrates made during photosynthesis into other compounds needed by living organisms (lipids, proteins, and nucleic acids), a source of minerals is needed. The role of carnivory is to supply these minerals. Instead of competing with other plants for the limited minerals of their habitat, carnivorous plants obtain these minerals from digestion of trapped animals.

The leaves of carnivorous plants have become modified to trap insects. So changed are these leaves that they only remotely resemble what one normally thinks of as leaves (although they still function in photosynthesis). Two major types of traps occur: passive and active. The passive traps do not employ any type of movement to entrap their prey, although ingenious methods have evolved that attract prey. The active traps not only attract prey but exhibit movement to catch and hold the prey. More detailed descriptions of these traps are given below.

Among the many myths associated with carnivorous plants is that some species are large enough to eat people. The truth is that the largest animal that could be caught by any carnivorous plant, anywhere in the world, is a small mouse—hence, the term carnivorous is preferred over insectivorous.

All carnivorous plants produce flowers and viable seeds. In the case of some pitcher-plants, the flowers develop before the leaves and one often does not equate the two structures as being from the same plant. The other four genera produce flowers and leaves (traps) at the same time.

Worldwide there are about 450 species of carnivorous plants representing fifteen genera. In South Carolina there are five genera, each of which, described below, occurs in the Lowcountry.

Pitcher-plants

Pitcher-plants have perennial rosettes of leaves modified to trap insects. The pitchers exhibit the passive trap and are tubular, appearing somewhat like elongated funnels, with a hood or lid at the top (plate 373). Usually the hood is supported by a narrow column and may be reflexed over the pitcher opening as in hooded pitcher-plant (plate 324) or be vertical as in frog's britches (plate 123). Two mechanisms attract insects: the coloration of the pitchers and the secretions of nectar from the nectar roll which is formed from the rolled up margin of the hood, the lip opposite the column, and the free margin of the wing. In frog's britches, once an insect lands on the brim of the opening or on the underside of the hood, downward pointing hairs on the lid force the insect into the pitcher. Often the insect loses its footing and falls into the pitcher. The upper portion of the pitcher is lined with a smooth wax which makes it difficult for the insect to crawl out. Ultimately the insect falls into the mixture of enzymes and fluids at the base of the pitcher where digestion takes place and absorption occurs through a wax-free zone.

The leaves of pitcher-plants arise from underground rhizomes in the early spring. It is these rhizomes that allow the plants to survive the spring fires that remove the competing vegetation. The rhizome also functions as a means of asexual reproduction.

Four species of pitcher-plants grow in the Lowcountry: hooded pitcher-plant (*Sarracenia minor,* plate 324), yellow trumpet (*S. flava,* plates 372 and 373), sweet pitcher-plant (*S. rubra,* plate 122), and frog's britches (*S. purpurea,* plate 123). The first two are common; sweet pitcher-plant is infrequent but widespread in longleaf pine savannas and pocosins; frog's britches is rare and found mainly in Horry and Georgetown counties in wet sphagnum openings in pocosins.

Sundews

Sundews are perennial (in our area) herbaceous plants with a basal rosette of leaves arising from a fibrous root system (plate 327). Leaves are produced continuously all year. A single stalk, arising from between the leaves, supports from five to twenty-five white flowers. The lowest flower opens for about two days, closes, and the next flower then opens. This process occurs throughout the spring and summer.

The leaves are an active trap to a limited degree and often referred to as a "flypaper" mechanism. The blade of the leaf is covered with stalked glands which secrete mucilage (which holds the insect) and digestive enzymes; the leaves then absorb the digested material. Small, crawling insects—either attracted to the plant by the coloration of the leaves, by the nectar from the glands, or just by chance wanderings—become mired in the sticky secretions. As the insect struggles to get free, by some yet unknown process, the motion caused by the insect sends signals through the blade to the long-stalked marginal glands which bend to the center of the leaf blade, further entangling the insect. In some species the leaf blade may become folded over the insect bringing a greater surface area for absorption in contact with the insect.

The common name of the genus, sundews, comes from the sticky, dewy tentacles that shine and glitter with the color of the sun's spectrum in the early morning. Three species occur in the Lowcountry and are common in longleaf pine flatwoods and savannas, in wet ditches, and along the edge of pocosins.

Butterworts

The butterworts' genus name, *Pinguicula*, is from the Latin word *pinguis,* meaning "fat" and the suffix *ula* meaning "little one." The leaves are greasy to the touch—hence, the genus name.

The butterworts are fibrous-rooted perennials that in the summer form a flat rosette of blunt, oblong leaves. The older leaves lie prostrate and the younger ones nearly so. The three species in the Lowcountry retain their leaves over the winter and do not form winter resting buds.

The leaf surface contains two kinds of glands: stalked glands that are important in catching and holding prey and sessile glands that are active in digestion. The trapping mechanism is the active flypaper type. Small prey, landing on the upper surface of the leaf, become mired in the secretions of the glands and are held until digestion and absorption occur. The leaves probably do not have any method of attracting prey since no nectar glands have been demonstrated. Crawling and flying insects probably come into contact with the leaves by chance. During summer, the active trapping season, the rolled edges tend to curl over the prey, perhaps keeping the partially digested insect from being washed off by rain.

The butterworts are very inconspicuous except when in flower and superficially resemble so many other plants in their rosette form that they are often overlooked. In late spring and early summer, however, the flowers make the butterworts highly conspicuous. Three species occur in the Lowcountry, one rare and two common. The two common species are yellow butterwort (*P. lutea,* plate 321) and violet butterwort (*P. caerulea,* plate 326). All three species occur in longleaf pine flatwoods and savannas.

Bladderworts

The common name, bladderworts, comes from the bladder-like structures, which are modified leaves, that function as traps. Species are either aquatic or terrestrial; the terrestrial species grow in moist soil.

The bladders are traps which are activated when an aquatic insect touches one of the sensitive hairs that surround the opening of the bladder. A hinged door quickly opens, water rushes into the bladder sweeping with it the insect and then closes—trapping the prey inside. Digestion takes place inside the bladder. Only small aquatic insects such as mosquito larvae can be trapped because of the small size of the bladders. The most prolific bladder producers are the aquatic species; terrestrial species only produce bladders when growing in very moist soil.

The most spectacular bladderwort is floating bladderwort (*Utricularia inflata* var. *inflata,* plate 108) which forms a remarkable flotation device that supports the flowering stalk. From six to ten floats (modified leaves) radiate from the middle of the stalk. Initial development of the stalk and bladders begins under water. As the floats grow, their buoyancy, caused by aerenchyma tissue, makes the entire plant rise to the surface. The stalk then produces the yellow flowers.

Three terrestrial and six aquatic species (including purple bladderwort, *U. purpurea,* plate 107) occur in the Lowcountry.

Venus' Fly Trap

Venus' fly trap (*Dionaea muscipula*) is a terrestrial carnivorous plant endemic to the Carolinas (plates 334 and 335). In South Carolina it occurs only in Georgetown and Horry counties. Good populations occur in the Lewis Ocean Bay Heritage Preserve and the Cartwheel Bay Heritage Preserve. It favors a damp, sandy soil with a small portion of peat (such as occurs around the Carolina bays); it is exacting in its requirements. Not yet near extinction, much of its habitat has been altered so that it must be monitored to ensure its preservation.

Adult plants have been successfully transplanted as far north as New Jersey. One population is still in New Jersey after thirty-plus years. Plants as far north as New Jersey do not reproduce sexually. Plants further south apparently can reproduce sexually, but the seedlings do not become established. The seedling stage apparently is the sensitive stage—the stage that shows the very specific habitat requirements.

The trapping mechanism, a modified leaf, is an active, springlike trap. The foliage is a rosette from a rhizome with each leaf blade of two lobes, each lobe attached to the midrib. Along the free margins of the lobes are pronglike teeth. On the surface of the lobes occurs two types of glands: nectar glands, near the margin to attract insects, and digestive glands. The latter turn red upon exposure to the sun, giving the leaf a red coloration which also attracts prey. On the upper surface of each are trigger hairs (generally three) which spring the trap when touched by an insect. The lobes close sufficiently in about half a second to trap the prey, after which digestion and absorption take place. To spring the trap requires two touches of a single hair or touches of two separate hairs, both within about 20 seconds. This prevents wind-blown objects from closing the trap and wasting energy. Usually each trap can catch three insects before it dies or becomes inactive. The continual production of new leaves during the growing season compensates for this loss. The white flowers are produced in the early summer.

PARASITIC PLANTS

Parasitism is a symbiotic relationship between two organisms living in direct contact; nutrition passes from one organism, the host, to the other organism, the parasite. In the relationship, the host is adversely affected to varying degrees, while the parasite benefits from the association. Parasitic relationships between animals are well known to laypeople and scientists. Who is not aware of tapeworms in animals as well as humans or had the displeasure of removing a tick after a trek through Lowcountry forests? Parasitism in plants, however, is equally common but not as well known.

Two major types of parasitic flowering plants evolved: hemiparasites and holoparasites. Hemiparasites contain chlorophyll and only depend on their hosts for water and minerals; they do, however, receive some organic material that is present in the xylem of the host. Mistletoe (*Phoradendron serotinum,* plate 136), dodder (*Cuscuta gronovii,* plate 92) and numerous genera of the Scrophulariaceae (figwort family) are hemiparasites. Holoparasites represent the most advanced condition in that they lack chlorophyll and do not conduct photosynthesis. They are dependent on their hosts for water, minerals, and food. Three holoparasites, all belonging to the broomrape family (Orobanchaeceae), occur in the Lowcountry: Cancer-root (*Conopholis americana,* plate 193), beech-drops (*Epifagus virginiana,* plate 209), and broomrape (*Orobanche uniflora*).

The one structure unique to parasitism in plants is the haustorium, a bridge of xylem tissue (figure 2) between host and parasite through which water, minerals, and limited amounts of food pass from the host to the parasite. The haustorium evolved from the undifferentiated xylem of the root of the parasite.

Reduction in seed size and leaf surface also occurred as parasitism evolved to the highly specialized holoparasite. Leaf surface became reduced because there was no selective pressure to maintain light-trapping structures. This reduction in leaf surface is clearly visible in cancer-root and beech-drops. In the holoparasites, which are parasitic of the roots of trees, seeds do not germinate until they get to the immediate rhizosphere (the soil immediately surrounding the root system of a plant). The seeds are carried down through the soil by water and, once in the rhizosphere, germinate only when stimulated by a chemical

Figure 2. Haustorium

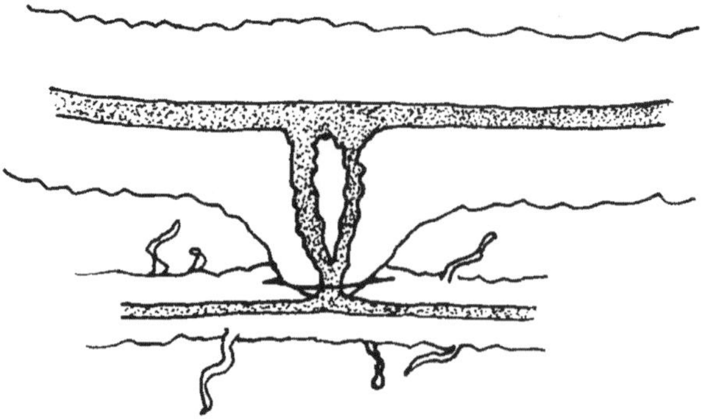

Host's Xylem

Haustorium

Parasite's Xylem

exudate from the roots. The selective advantage of the small seeds is obvious: small seeds are able to pass through the pores in the soil more readily, and their size makes it possible to produce vast numbers per capsule. After germination the embryonic root of the seed makes contact with the host root and adheres to it by a special tissue. An intrusive organ develops that penetrates the root and then the bridge of xylem forms. This combination of adhesive tissue, intrusive organ, and xylem functions as the haustorium.

There is great variation in the effects of the parasite's interference in the physiology of the host, ranging from spectacular malformations to absence of recognizable symptoms. A heavy infestation of mistletoes on fruit trees may cause a significant decrease in the reproduction capacity of the host trees and result in a reduced fruit crop. The dodders are noxious weeds on crops such as clover and alfalfa, and in a very short time a crop can be covered by a thick web of filamentous dodder stems that drain minerals from the host plant. On the other hand, beech-drops and cancer-root, their mass small when compared to the trees they parasitize, have little effect on their hosts. The parasite does not usually kill the host plant because the life-support system of the parasite would be eliminated at the same time. Parasitism would never have evolved if the host were killed. The system has evolved to the "prudent" point that the host generally suffers some debilitating effects but does not die. The parasite is thus ensured of its life-support system.

The most spectacular of the parasitic flowering plants are members of the figwort family. Seven genera are represented in the local flora; six are represented in this book: *Aureolaria, Agalinis, Pedicularis, Schwalbea, Buchnera,* and *Castilleja*. The parasitic figworts all have brightly colored flowers and are hemiparasites. There is little precise information on the specific hosts of these parasitic figworts. There is no disagreement, however, that they are spectacular wildflowers.

SAPROPHYTIC PLANTS AND MYCOTROPHY

Plants that live by using preformed organic material (food) are heterotrophic, which means they have no chlorophyll and cannot carry on photosynthesis. Their food source comes from one of two sources: they are either parasitic and obtain their food from a living host or they are saprophytic and obtain their food from decaying organic matter. In the plant world the majority of saprophytes are the fungi and bacteria. These decomposers are essential to the decay of biological wastes and dead organisms by returning the carbon of these wastes to the air as carbon dioxide and releasing minerals to the soil to be used again.

Saprophytism is not confined to the lower plants. Two families of flowering plants, the Orchidaceae and the Ericaceae (heath family, to which the azaleas belong), have saprophytic representatives. Saprophytism in flowering plants probably evolved in photosynthetic species that grew in forests. A mechanism first evolved to transfer food from the soil environment into the photosynthetic plant; then, with the evolutionary loss of the ability to make chlorophyll, the ready source of food in the forest environment allowed for the evolution into saprophytism. Flowering saprophytes are still confined to forests because a build-up of dead organic matter occurs in this habitat.

The mechanism that evolved to transfer food from the soil into the saprophyte is mycotrophy (the Greek word *mykes* means "fungus," *trophia* means "nourishment"). The roots of many photosynthetic vascular plants harbor a filamentous fungus. The roots and fungus form a combination called a mycorrhizae. The role of the fungus is to absorb water and minerals from the soil and pass it to the host plant. Mycorrhizal absorption of water and minerals is often more efficient than that which would be carried on by the unaided root because of the increased absorbing surface provided by the fungal hyphae. Some biologists believe that the additional water provided by the fungus is necessary under certain natural conditions to supply water for many forest trees to replace that lost by transpiration. The mycorrhizal fungus can also tap a supply of minerals that are not readily available to the vascular plant from its digestion of organic matter in the soil.

The mycorrhizal association between the saprophytic flowering plants probably evolved from a mycorrhizal association of fungi and photosynthetic plants. With the loss of the ability to make chlorophyll, the fungus became the vehicle to pass not only water and minerals from the soil to the saprophyte, but also food. The relationship is obligate, and the saprophyte will not survive without the fungus.

A new theory has surfaced among scientists concerning the source of food transferred to the saprophyte by the fungus. The older view is that the food comes from the breakdown of organic matter in the soil by the fungus. Recent studies, however, suggest a more evolved system. The mycorrhizal fungus may be associated with a second plant—a green, actively photosynthetic angiosperm. The fungus forms a bridge that transfers carbohydrates from the green plant to the saprophyte. According to this theory, the saprophytes are apparently indirect parasites (epiparasites) of photosynthetic plants.

Five species of saprophytic flowering plants occur locally—two members of the Ericaceae family: Indian pipe (*Monotropa uniflora,* plate 235) and pine-sap (*M. hypopithys*); plus three members of the orchid family: autumn coral-root, spring coral-root, and crested coral-root.

EPIPHYTIC PLANTS

Epiphytes grow on other plants. The term is derived from the Greek words *epi* meaning "upon" and *phyte* meaning "plant." The evolution of the epiphytic habitat allowed these plants to raise themselves above the competition of the forest floor and move into a habitat of lesser competition. Epiphytes are distinguished from parasites in that epiphytes depend on their host only for physical support. Epiphytes include a variety of plants such as fungi, algae, mosses, liverworts, lichens, and vascular plants. Three vascular epiphytes occur in the Lowcountry: resurrection fern (*Polypodium polypodioides,* plate 32), green-fly orchid (*Epidendrum conopseum,* plate 153), and Spanish moss (*Tillandsia usneoides,* plate 33) (the latter two are flowering plants).

Of all ecological groups of plants, epiphytes are the most dependent on rain for their water supply; their survival depends on being able to endure drought. Epiphytes evolved specialized structures that absorb moisture from the atmosphere. In Spanish moss the roots serve mainly as anchorage while the stems and leaves have taken over the role of absorption. The leaves and stems are covered with gray, peltate scales that collect and absorb water by capillary action. The water is then absorbed through uncutinized areas on the stem and leaf. During drought, when the plant shrinks in size, scales cover the uncutinized areas and prevent water loss. In the tropical orchids, a development of empty, whitish cells from the epidermis of the root, called the velamen, can take up water rapidly from the atmosphere even after a light rain. The living core of the root then absorbs the water from this storage tissue.

Vascular epiphytes are photosynthetic plants. Both carbon dioxide and water needed for photosynthesis are obtained from the atmosphere. Minerals are derived from three sources: (1) decaying bark of the host tree, (2) rainwater which collects dissolved substances, and (3) wind-borne particles which collect in the crevices of the bark of the host tree. Once the plant produces the basic carbohydrate in photosynthesis, it can combine the carbohydrate with the minerals to make other organic compounds necessary to survive.

Epiphytes and their hosts are well-adapted to their relationship and the epiphytes cause no appreciable damage to their hosts. Occasionally, however, the weight of the epiphyte may break a supporting branch and allow fungi and insects to infect the host or massive growths of Spanish moss may block sunlight to the branches and reduce normal foliage. Spanish moss has been attributed to reducing the yield of pecan trees because of the shadowing of young buds.

Vascular epiphytes in the Lowcountry are generally located in the forks of trees and on horizontal branches; they are less common on vertical and smooth surfaces. Horizontal branches and forks provide a place where organic matter can accumulate thereby creating conditions more conducive for establishment and maintenance of propagules.

NATIVE ORCHIDS

The orchid family (Orchidaceae) is one of the most fascinating and diverse families of flowering plants in the world. A cosmopolitan family attaining its highest development in the mountains of the tropics and subtropics of both hemispheres, the orchid family is found throughout the world where some amount of moisture is present. Orchids are absent only from absolute deserts and from polar regions where the ground is permanently frozen. There are approximately thirty thousand species of orchids worldwide. Within the United States and Canada, botanists recognize 210 species and varieties, and within South Carolina, forty-five species and varieties and two hybrids have been identified. Thirty-eight species, two varieties, and one hybrid occur in the Lowcountry.

From an artistic and aesthetic viewpoint, orchids are universally accorded first place in nature. The beauty of their flowers has made the orchid family the center of a multimillion-dollar floral industry in the United States, but the family is otherwise of little economic importance. The most important product is natural vanilla from the seed pods of a species grown in Madagascar.

Despite its wide range of habitat and large numbers, the orchid family is often overlooked by botanists and laypeople because most associate orchids with exotic tropical species or with the myriads of large-flowered hybrids sold for corsages. In fact, the majority of our native orchids have small, inconspicuous flowers. On close examination, however, the native orchids are just as alluring and attractive.

Orchids are extremely specialized perennial herbs that are either terrestrial, lithophytic (growing on rocks), or epiphytic (growing on another plant, usually a tree). Terrestrial orchids are common in temperate areas such as South Carolina, whereas epiphytic orchids are most common in tropical areas. With the exception of the green-fly orchid, all orchids in South Carolina are terrestrial; no lithophytic orchids occur in South Carolina.

In the tropics and subtropics, the orchid's epiphytic growth habit evolved as an adaptation to sites where there was virtually no sunlight at ground level. By growing on the trunks or highest branches of trees, where sunlight was abundant, orchids' photosynthetic organs were raised above the competition from plants of the forest floor. In South Carolina the green-fly orchid (plate 153) is the only epiphytic representative. The epiphytic habitat is restricted north of the subtropics because of both the cold and insufficient moisture. Exposed to the atmosphere, and lacking the insulation of the soil, the roots of epiphytes are killed in cold temperatures. Only along the coast of South Carolina, where the ocean keeps the temperature a few degrees above the inland areas and where moisture is higher, is the epiphytic lifestyle possible for orchids. Most commonly the green-fly orchid is found on the branches of tall cypress and gum trees in swamp forests or on live oaks in upland habitats.

Many orchids, after a normal period of flowering, do not produce above-ground vegetative parts for a year or more. In other words, they disappear for a time. A bog covered with snowy orchid (*Habenaria nivea,* plate 341) one year may show no signs for years, only to return in great numbers another year. Some reports claim that the orchid is building up food supplies in the underground stem during the hiatus in order to provide energy for the flowering process.

One of the most unusual lifestyles of specialized orchids is saprophytism. Saprophytes are plants that lack chlorophyll and are unable to manufacture food. They obtain food from decaying organic matter in the soil. The saprophytic orchids are terrestrial species that depend on a soil fungus to supply the food. The fungus lives in the roots of the orchid and sends out its mycelium into the soil. This mass of mycelium greatly increases the surface area to absorb water, minerals, and food. The fungus secretes enzymes into the soil to digest the organic material, then absorbs the digested, simpler compounds and transfers them to the orchid. Usually saprophytic orchids are confined to humus- and moisture-rich habitats.

Three species of saprophytic orchids occur in the Lowcountry: autumn coral-root (*Corallorhiza odontorhiza*), spring coral-root (*C. wisteriana*), and crested coral-root (*Hexalectris spicata,* plate 180).

The orchid family is also unusual among higher plants for three reasons: (1) the variation in its highly specialized flowers, (2) the large number of seeds without endosperm in each capsule, and (3) its great diversity of habitat. The flower is unusual in that the male and female parts (stamens and pistil) are fused into one structure called the column. At the apex of the column is the male anther with its pollen grains grouped into masses called pollinia. Below the anther is the stigma, the terminal portion of the pistil. The surface of the stigma is sticky so that pollinia brought by insects from other flowers will adhere to the stigmatic surface. Orchids are pollinated primarily by insects, with each species appearing to have its own insect species to effect pollination. All orchid flowers have three sepals (outer whorl) and three petals (inner whorl). One of the petals, the labellum or lip, is different from the other two—often larger and more showy. Usually the flower grows so that the lip is lowermost; however, in the genus *Calopogon* (plates 320 and 331) the lip is uppermost.

The seed capsule of orchids requires about nine months to mature and may contain millions of dustlike seeds. The seeds contain no endosperm and are dependent on an external aid for germination and growth of the seedling. In nature a soil fungus penetrates the seed and establishes a symbiotic relationship with the embryo. The embryo grows by absorbing the waste or by-products produced externally through digestion or secretion by the fungus. The fungus will eventually penetrate the roots of the orchid where it remains in a symbiotic relationship (mycorrhizal) throughout the orchid's life. Thus, all orchids, at least as seedlings, are saprophytic. The delicate balance of orchid, soil, and soil fungus is such that transplanting orchids from their native habitat is difficult.

The diversity of habitat and growth forms of orchids attests to the remarkable adaptation of the orchid family. In the Lowcountry, two species of *Spiranthes* have been found in abandoned farm fields and another species in freshwater pools behind coastal dunes. Five species of *Habenaria* occur in longleaf pine savannas while two other species of *Habenaria* occur in swamp forests. In the freshwater tidal marshes along the coastal rivers there are two species: water-spider orchid (*Habenaria repens,* plate 69) and fragrant ladies' tresses (*Spiranthes odorata,* plate 91). The greatest number of species, however, occurs in the beech forests where nine different orchids grace this community. Furthermore, some of the orchids found in the Lowcountry grow statewide, while others only grow in restricted habitats within the Lowcountry.

WOODY VINES

Woody vines represent a specialized evolutionary adaptation in that they make use of other plants' supporting tissue to allow their leaves to get above competing plants' leaves where they receive more sunlight. By definition, woody vines are vascular plants that remain rooted in the soil but keep their stems erect by use of other objects for support. Supporting objects are other plants or structures such as telephone poles, fence posts, and buildings. Woody vines are not parasitic since they obtain no food from the host, only support. They can be harmful, nevertheless, in two ways to supporting trees and shrubs: (1) by spreading their foliage over the host plant and blocking its source of sunlight, and (2) by twining and constricting the supporting hosts' trunk or stem such that downward translocation of food may be impaired which in extreme cases can kill the tree.

Woody vines have little economic importance in the United States with the exception of the genus *Vitis* (grape). They do provide, however, valuable forage, seeds, and fruits for wildlife. In the tropics, woody vines are important in a negative manner to the forestry industry. Often trees may be woven together by numerous vines so that a single tree cannot be felled, and trees must be cut in groups instead.

Woody vines that occur in the Lowcountry can be classified into four categories based on their method of climbing.

Thorns

These vines have thorns or prickles that act in a passive manner to aid in climbing. Cherokee rose (*Rosa laevigata,* plate 397) is an example of this type. Its clinging lifeform is undoubtedly aided by its large, curved prickles.

Twiners

Twiners are woody vines in which the entire stem twines about the supporting structure. An unusual feature of twiners is that within a particular taxon, twining is either consistently clockwise or consistently counter-clockwise.

Numerous species of twining vines occur in the coastal area: supplejack, Japanese honeysuckle, coral honeysuckle, yellow jessamine, native wisteria, and Chinese wisteria.

Tendrils

Tendrils are specialized organs to facilitate climbing and may be modified parts of leaves, stipules, leaf stalks, roots, or stems. The tendrils may either twine about the supporting structure or adhere to the surface of the support. Initially tendrils are weak structures, but once they become attached to the supporting structure their mechanical tissue develops to such a degree that it is capable of supporting the weight of the vine.

Numerous vines that climb by tendrils have showy flowers (or fruits) and are considered wildflowers. Representatives in the Lowcountry with conspicuous flowers or fruits include Virginia creeper, poison ivy, cross vine, cow-itch, climbing hydrangea, coral greenbrier, and bamboo-vine.

Flattened Stems

Climbing fetterbush (*Pieris phillyreifolia,* plate 385) is a woody vine with an unusual habit of climbing. The vine grows under or in the crevices of the outer bark of trees. The host tree is pond cypress (*Taxodium ascendens*). The plant originates as a shrub at the base of the host tree; in time its main branch may grow under or in the crevice of the outer bark. When it does, the stem assumes a flattened form which better enables it to grow under the bark. At a place considerably higher up the tree, the stem sends out lateral branches. These lateral branches produce white flowers in the spring.

DUCKWEEDS

The members of the duckweed family, Lemnaceae, are the smallest of all flowering plants. Members range in size from *Wolffia,* the smallest flowering plant barley discernable to the naked eye, to *Spirodela polyrrhiza,* the largest of the duckweeds (fronds about .02"). Duckweeds are aquatic plants free-floating on or below the water surface. Aquatic habitats include sluggish fresh waters of ponds, pools, lakes, swamps, streams, drainage ditches, canals, and sloughs. Acidic waters with high organic content seem to favor duckweeds, although they are also found in crystal clear waters of springs.

Structurally the plants consist of a frond, or thallus, and are not differentiated into stems or leaves. The frond acts as a leaf. Asexual reproduction occurs by budding from the parent plant, with the daughter plant then separating from the parent or remaining attached by a short stipe thereby producing connected colonies. Sexual reproduction by flowers occurs infrequently; flowers are produced in small pouches, in pits at the edge of the frond, or on its upper surface. The flowers are too small to be seen with the naked eye.

A species of duckweed may be present in small numbers in a given body of water one day, but in great abundance in the same place at another time. The several kinds may be intermixed or may occur singly. Often only the edge of a pond is populated followed by a bloom of a few weeks duration in which the surface becomes covered by a dense growth. Within another few weeks most of the growth disappears from the open water and only the edge is populated. When a body of water dries up, the plants are often stranded in mats along the edge. As long as the substratum remains wet, the plants remain alive and can repopulate the system when it fills with water.

Duckweeds are used as food for a variety of waterfowl. Aquatic animals associated with the duckweeds probably contribute significantly to the food value of the diet.

Four genera of duckweeds occur in the Lowcountry: *Spirodela* (duckmeat), *Lemna* (duckweed), *Wolffia* (water-meal), and *Wolffiella* (bog-mat). Plate 104 depicts a bloom of *Spirodela* and *Wolffiella* on a Lowcountry pond. The four genera are easily recognized with the naked eye or hand lens. *Spirodela* has several roots per frond and is red beneath, *Lemna* has but one root per frond and is green beneath, *Wolffia* consists of almost meal-like bodies without roots, and *Wolffiella* is made up of straplike bodies without roots. The following diagrams in figure 3 of the four genera will aid identification.

Figure 3.
Genera of Lemnaceae

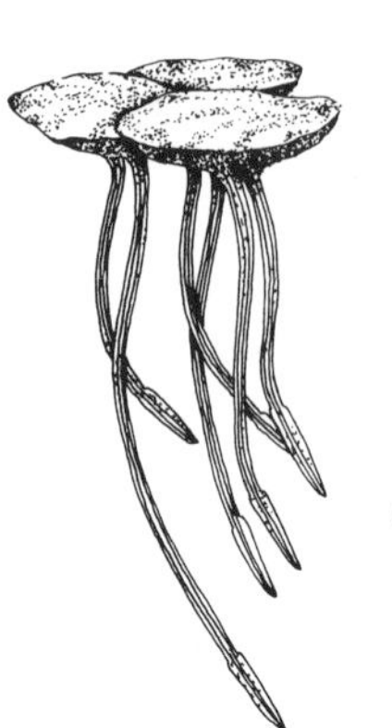

a. Duckmeat, *Spirodela polyrhiza*

b. Bog-mat, *Wolffiella floridana*

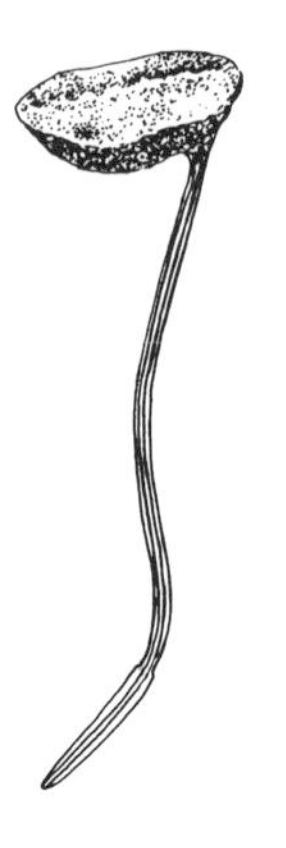

c. Duckweed, *Lemna valdiviana*

d. Water-meal, *Wolffia papulifera*

RARE PLANTS

A rare species is difficult to define since the concept involves two variables: the overall distribution and the abundance or density within that distribution. Figure 4 (from Hardin, 1977) illustrates this relationship. The extreme in rareness, Point A, would be a species like Carolina spleenwort (*Asplenium heteroresiliens*), represented by few individual plants, but also restricted to a extremely limited geographical area. Point B is represented by American chaff-seed (*Schwalbea americana,* plate 291) which occurs over a fairly broad geographical area, but at very low density. Point C is represented by pond spice (*Litsea aestivalis*), which is locally abundant in the Francis Marion National Forest but still rare since its total distribution is limited. The concept of rareness, then, includes all possible intermediates among these three extremes.

The Endangered Species Act of 1966 first recognized the intrinsic value of species and noted that some species were so rare that they were on the verge of extinction. The 1966 act only recognized animals; in 1973 the act was broadened to include plants. Since the passage of these acts, numerous organizations have come forward with programs and methods to locate and provide protected habitats for listed species. But perhaps more important is the realization by these groups that preservation of our natural heritage depends on preservation of natural diversity. With this concept as its standard, various organizations have protected numerous sites in the Lowcountry. Many of these sites are listed in part 3 since they harbor examples of the natural gardens described in this book.

Several categories of plant rarity are recognized, with the first two given legal status by the U.S. Department of Interior under the Endangered Species Act.

An endangered species is one that is in danger of becoming

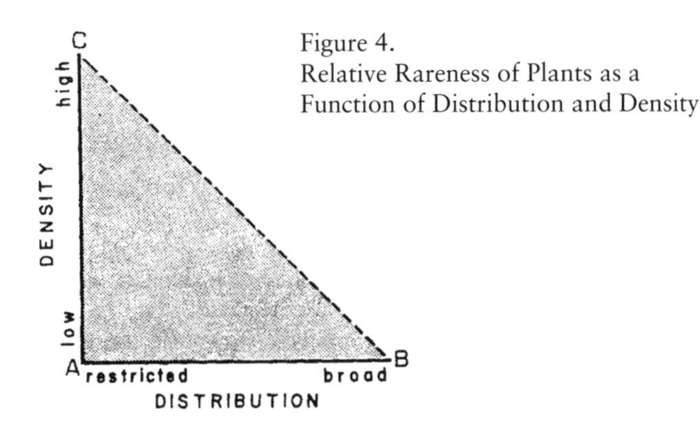

Figure 4.
Relative Rareness of Plants as a Function of Distribution and Density

extinct in the near future. In the Lowcountry, only four species have been listed as endangered: pond berry (*Lindera melissifolia,* plates 383 and 384), American chaff-seed (*Schwalbea americana,* plate 291), Canby's dropwort (*Oxypolis canbyi*), and sea-beach amaranth (*Amaranthus pumilus*).

A threatened species is one not now endangered but is heading in that direction. No threatened species are included in this book since none are listed for the Lowcountry.

A peripheral species is one that may be common in adjacent states, but rare at the edge of its distribution in South Carolina. An example is fever tree (*Pinckneya pubens,* plate 169) which grows throughout the coastal plain of Georgia but is known from only two locations in South Carolina—one in Beaufort County and one in Jasper County. State botanists consider this a species of special concern.

A disjunct population is one separated by a substantial distance from its main area of occurrence. An example is Indian paint brush (*Castilleja coccinea,* plate 325) which is found primarily in the mountains and piedmont, but also occurs in Berkeley and Williamsburg counties in the Lowcountry.

An endemic species is one with its native habitat confined to a small area of a state or states. The Venus' fly trap is an example as it is confined to the southeast coastal plain of North Carolina and Horry and Georgetown counties in South Carolina.

NATURALIZED PLANTS

The majority of flowering plants in disturbed sites are nonnative species that have replaced or supplemented our native wildflowers. One only has to look at the section on the ruderal gardens to see how many of the included species are aliens that followed the early settlers across the Atlantic from Europe (and to a lesser extent from other parts of the world). The list of naturalized plants is long: henbit, crimson clover, sour clover, common dandelion, wild carrot, yarrow, prickly poppy, and wooly mullein, to name but a few. These species, which easily adapt to newly created or disturbed sites, quickly found a home; that is, they became naturalized because they could reproduce, become established, and spread without cultivation. Approximately 10 percent of South Carolina's present flora is nonnative.

The coastal area of South Carolina is home to hundreds of alien plants. This is easily understood when you consider the first cities were seaports along the coast (for example, Savannah, Beaufort, Charleston, and Georgetown). The ports were sites of entry for these alien species that came mixed with shipments from across the sea. Once established around the port cities, their propagules were unwittingly carried into the interior. As the native habitats became altered, the alien species gained additional footholds. Drainage ditches lowered water tables and changed sloughs and wetlands to dry lands. Woods were cleared for fields and homesites. Railroads bridged rivers that might have created a barrier, if only temporarily, for the spread of these alien species. Hardwood and pine forests were often clear-cut of timber, creating additional disturbed ground. One hundred-fifty thousand acres of tidal freshwater swamp forests along the coastal rivers were banked and then cleared of vegetation to grow rice. After the industry collapsed in the early 1900s, the fields were abandoned—creating ideal sites for alien, aquatic weeds.

In modern times, with the invention of electricity, inland lakes were created to provide water power to generate electricity. Free-flowing rivers were converted into lakes—ideal sites for exotic, aquatic weeds such as water hyacinth, water primrose, and alligator-weed.

A grass that is causing a serious problem in coastal marshes is common reed (*Phragmites australis,* plate 455). It reproduces rapidly by rhizomes and long, leafy stolons and can quickly dominate a marsh to the exclusion of all native species. It first began to appear in the South Carolina coastal area in the mid-1970s in dredge soil disposal sites along the Intracoastal Waterway. Some scientists believe it spreads by its rhizomes being embedded in mud of dredges which move from site to site along the Waterway. Because it has little food value for wildlife, its replacement of the native species is a serious problem for wildlife managers.

Not all the alien species entered accidentally. Numerous species were brought from overseas to adorn gardens of plantations. André Michaux introduced mimosa, popcorn tree, and China-berry shortly after he established his botanical garden in 1786; all three have became naturalized and are today considered common. Rattlebox (*Crotalaria spectabilis,* plate 453), brought from Japan, was cultivated as a soil-building, green manure. It escaped and has become one of our most noxious weeds in agricultural fields. The legendary kudzu vine, introduced from Japan for erosion control, escaped and is a major pest in fields and woodlands. Its dense growth blocks sunlight from the leaves of the tree on which it grows, often killing the tree. *Lespedeza bicolor* (plate 305), commonly called bicolor, is a legume introduced from Japan by wildlife technicians as a winter food for quail. It was planted in strips or patches on many game preserves and has now spread into surrounding woodlands. White mulberry (*Morus alba*) was introduced by the early colonists for the silk industry which never materialized. The tree, however, became naturalized and often dominates disturbed areas such as dredge soil disposal sites (for example, Drum Island in Charleston Harbor).

Another event that might further shift the balance between alien and native species today is the program of planting wildflower strips along the major roadways and vacant lots in the cities. This is an ambitious program and the results are spectacular. How many of these species, if any, will become naturalized waits to be seen.

EDIBLE WILD PLANTS

The coastal area of South Carolina is a botanical paradise for anyone who wishes to become independent of the conventional sources of food, or anyone who simply wishes to sample wild plants in a limited manner. Numerous species of native and naturalized plants in the Lowcountry are sources of beverages or food, many of which are described in this book.

Almost anyone with a limited exposure to the outdoors is familiar with some of the common edible plants such as plums, blackberries, hickory nuts, blueberries, and grapes. But only the true outdoor person knows the great variety of edible wild plants. Many people hesitate to seek wild foods because there are so many poisonous plants. The poisonous plants are easily identified by a careful observer who takes time to study them. The best recommendation for someone who is serious about learning the edible and poisonous plants is to take an introductory course in plant taxonomy or field botany. But for those unable to do so, frequent field trips with persons knowledgeable of the local plants will suffice.

Below are various categories of edible wild plants. Note that some species have parts that can be included in more than one category. There is no attempt to give recipes for the plants; one need only to refer to the books in the bibliography for this information.

Flour

A large number of plants can be used as breadstuffs or be ground into flour. Wildflowers depicted in this book that can be made into flours include cat-tail, ground-nut, and soft-stem bulrush.

Teas

The forests and fields abound with plants that can be used as substitutes for Oriental teas. Many of the plants used as teas have a history of medicinal use. Others, however, are used because they possess a pleasant flavor. Some came into use during the American Revolution when Oriental tea was under embargo. Whatever their origin, tea substitutes have grown in popularity in recent years and are now widely used. Native and naturalized plants used for teas include sassafras, witch-hazel, dewberry, yaupon, New Jersey tea, and wax myrtle.

Pickles

A variety of native plants have been used throughout America as pickles. The plant part is first soaked in alum-water, then in salted water, and finally preserved by boiling in spiced vinegar. Roots, tubers, leafy young plants, flower-buds, and young fruits all have been pickled. Wildflowers that can be pickled include cat-tail, young pokeweed leaves, elderberry flowers and redbud flowers.

Fritters

Fritters are prepared by cutting flower blossoms at the peak of bloom and soaking in brandy with a stick of cinnamon for an hour. Each flower or flower cluster is then dipped into an egg batter and dropped into hot fat until fried to a light brown. The fritters may be sprinkled with a variety of seasonings. Wildflowers that have commonly been used as fritters include elderberry, cat-tail, day-lily, and common dandelion.

Jellies and Marmalades

The local flora abounds with plants whose fruits can be made into jellies and marmalades and include may-apple, red chokeberry, swamp rose, crab-apple, chickasaw plum, muscadine, persimmon, elderberry, blueberries, huckleberry, dewberries, and viburnums.

Condiments and Seasonings

Numerous wildflowers can be used as substitutes for traditional condiments and seasonings. These include wax myrtle, sassafras, swamp rose, day-lily, elderberry, wild ginger, and red bay.

Rennet Substitutes

Warm milk can be turned into a curd by pouring it into fruit syrups just like it is curdled in the kitchen by using rennets. Besides the fruit-acids, several wild plants that can substitute for traditional rennets include species of butterworts and sundews.

Wines and Beers

Many plants exist that have been made into beer or wine at some point in history. While all plant parts have been used, those made from the flowers and fruits are more suitable.

Local plants used for wines include dewberry, blackberry, black cherry, common dandelion, elderberry, muscadine, and chickasaw plum; for beer, persimmon and catbrier.

Starchy Soups

Starchy soups are made from substances which, when boiled, both thicken and add nutrients to a soup. Three wildflowers can be used in this way: blue violets, sassafras, and day-lily.

Fresh Fruits

Wildflowers producing edible fruits include may-apple, blackberries, dewberries, chickasaw plum, muscadine, maypops, elderberry, huckleberries, blueberries, ground-cherries, and partridge berry.

Nuts and Large Seeds

Mockernut hickory produces edible nuts, while pickerelweed and sacred bean produce edible, large seeds.

Coffee Substitutes

In the local flora there are three caffeine-free species used as substitutes for coffee: common dandelion roots, persimmon fruits, and acorns.

Cooked Green Vegetables

Numerous wildflowers can be cooked and served like familiar garden vegetables. These include cat-tail, common dandelion, day-lily, pokeweed, young shoots of *Smilax* ssp., and redbud.

Cold Drinks

In this group are acid fruits that can be used to make a cold drink. Examples include blackberries, dewberries, black cherry, winged sumac, elderberry, muscadine, red maple, sassafras, and maypops.

Salads

The following plants contain tender parts that can be eaten without cooking, including ones that can be used in place of lettuce or even eaten raw. Examples are young shoots of *Smilax* ssp., cat-tail, chickweed, common dandelion, day-lily, violets, redbud, pickerelweed, and prickly sow-thistle.

Masticatories

Masticatories are plants that are chewed to relieve thirst. Some of the lesser known masticatories are soft-stem bulrush, sourgrass, horse sugar, and muscadine.

MEDICINAL PLANTS AND FOLK REMEDIES

From earliest times people have used plants in an attempt to cure diseases and relieve pain. These primitive attempts at medicine were often based on speculation and superstition that diseases were due to evil spirits in the body and could only be expelled by ingestion of substances disagreeable to these spirits. Curative agents must have been discovered by trial and error, and the knowledge accumulated slowly and spread by word of mouth from generation to generation.

Perhaps the first substantial record of plants in medicine comes from the Code of Hammurabi, a series of tablets from Babylon from about 1770 B.C. It mentions such plants as henbane, licorice, and mint which are still used in medicine. The Egyptians recorded their cures on temple walls and on the *Ebers Papyrus* (1550 B.C.) which contains over seven hundred medicinal formulas—including substances from species now known to have therapeutic value, such as castor-oil plant, mandrake, and hemp.

The Golden Age of Greece was a time of great advancement in medicinal and biological knowledge. Men such as Hippocrates, Aristotle, and Theophrastus contributed much to the science of pharmacology. Hippocrates did not believe evil spirits caused sickness and began to prescribe plant products as cures. The most significant Greek work was *De materia medica* by Dioscorides in 77 B.C. which dealt with the nature and properties of all the medicinal substances known at that time. Although poorly organized and often inaccurate, it became the prototype for future pharmacopoeias and was accepted without question by Europeans until the fifteenth century.

The Middle Ages was a period of relative stagnation in medicine. Some historians blame this stagnation on Dioscorides and his contemporaries because they claim Europeans blindly followed detailed but often incorrect ideas about diseases and their cures. Others blame it on the dormancy of intellectualism in Europe. But some progress was made during this period as botany and medicine became more closely linked.

In the fourteenth century, the Renaissance brought with it a new desire for knowledge. Studies of the human body were renewed and surgical procedures improved. Medicine did suffer a slight setback when Paracelsus (1493–1541) denounced the works of the Greeks and proposed the Doctrine of Signatures. According to this superstitious doctrine, all plants possessed some sign given by the Creator which indicated the use for which they were intended. Thus a plant with heart-shaped leaves should be used for heart ailments. As absurd as it seems now, it received great acclaim when proposed. It was soon displaced by less subjective and more secular methods of determining a plant's medicinal properties.

By the seventeenth and eighteenth centuries, science advanced to the testing of hypotheses which led to an improved understanding of physiology and provided a framework for testing medicines. The first half of the twentieth century saw advancements in medicine as causes of diseases were discovered and new drugs were isolated and synthesized. Many of these drugs were products from traditional, plant-derived extracts such as morphine, digitalis, quinine, and ephedrine. Today synthetic drugs are used more because of the cost of bringing a new plant-derived drug to market. Nevertheless, plants are still being investigated as possible wonder drugs, especially as potential antitumor agents.

The history of medicinal plants and folk remedies in the Lowcountry is a microcosm of the history of medicinal drugs as outlined above. Long before the settlers arrived in the New World and through generations of trial and error, Native Americans learned well the medicinal uses of the plants growing around them. Indeed, the highly developed cultures of the Aztecs, Mayans, and Incas may have been in part attributable to their understanding of the value of plants as medicines. So complete was the aboriginal's knowledge of the native flora that Indian usage can be demonstrated for all but a few of our indigenous vegetable drugs. In fact, many of the plants used by the Indians have been included in various formularies and pharmacopoeias in the United States and elsewhere. Extracts of the roots of may-apple (*Podophyllum peltatum,* plate 189) were used by the Indians as purgatives and for skin disorders and tumorous growth; today, may-apple alkaloids are used to treat lymphocytic leukemia. Indians used the bark of the willow tree (*Salix* ssp.) to relieve headaches and ease sore muscles. The active compound in willow bark is salicin. Today a derivative of this, acetylsalicylic acid, is known as aspirin. New Jersey tea (*Ceanothus americanus,* plate 225) was a favorite tea substitute during the American Revolution among the settlers who noted that the Indians used it frequently. Research in Europe has shown the plant has a potential for reducing high blood pressure.

In the 1600s, settlers from Europe began the trek from the Old World to the New World. Along the coast of South Carolina a plantation system developed—first based on rice, then indigo (briefly), and then cotton. The slaves, as well as the colonists, quickly acquired the knowledge of the Indians' use of native plants as folk remedies. The settlers also had access to European medical books and herbals. By the mid-nineteenth century the medicine of the colonists contained many elements of Native American and European medicine. In addition, the slaves, who came from cultures that employed plants for remedies, began a discovery of plants in the forests and fields which could be used medicinally.

The development of folk medicines by plantation slaves was a necessity. Plantations in the 1700s and 1800s were large and located along the rivers or on the sea islands such as Edisto Island or James Island. Communication between all but the adjacent plantations was difficult; trips to cities were even more difficult. Planters who maintained satisfactory medical facilities for the treatment of diseases were the exception. Those who did attempt to care for the slaves were fairly ignorant of the importance of good health measures. Except in the most severe cases, these masters and overseers made their own diagnoses and prescribed remedies without the aid of a doctor, who was infrequently employed. In addition, the planters generally left the plantations in the summer and moved to the pineland villages or ocean villages to escape the ravages of malaria, leaving the care of the slaves to overseers. More often than not the overseer was unable or chose not to adequately administer the medical needs of the slaves. Under these conditions, the slaves maintained their own medical practices, especially to cure everyday illnesses.

The use of native folk remedies was not confined to the slaves or planters. Most people living in the 1700s and 1800s in rural South Carolina made use of folk remedies. During the Civil War when the South was cut off from imported drugs because of the naval blockade, it turned to its native resources for drugs and food. The surgeon general of the Confederate Army commissioned Dr. Francis Peyer Porcher of St. John's Parish, Berkeley County, to explore the possibilities of using native plants of the South in order that, as Dr. Porcher stated, "the physician in his private practice, the planter on his estate, or . . . the regimental surgeon in the field may collect these substances within reach, which are frequently quite as valuable as others obtained abroad." Within a year's time, Dr. Porcher published in 1863 his extraordinary and classic book, *Resources of the Southern Fields and Forest,* in which he listed over six hundred species of plants that were of some value to the South. So complete was this book, revised and enlarged in 1869, that it is still referred to today.

Dr. Porcher also gives an interesting account of Jimson-weed (*Datura stramonium,* plate 452), the only hallucinogenic flowering plant in the Lowcountry. By definition, hallucinogens are nonaddictive substances which in nontoxic doses cause temporary changes in perception, mood, and thought with a distorted sense of reality. Jimson-weed is an abbreviation of Jamestown-weed—a name acquired when British troops landed in Jamestown Colony to put down a tax rebellion. Inadvertently, the leaves of *Datura* were included in their meal, and, as reported in Robert Beverly's *History of Virginia,* "the effect of which was a very pleasant Comedy; for they turn'd natural Fools upon it for several days . . . and after Eleven Days, returned to themselves again, not remembering any thing that pass'd." Since the plant contains numerous alkaloids, it is also highly poisonous.

The use of folk remedies has not ceased in modern, rural South Carolina. Dr. Julia Morton, in her 1974 book *Folk Remedies of the Low Country,* documented that rural people today still use the local plants for a variety of folk remedies. Some of the uses can be attributed to the recent "back to nature" fad as people turn away from refined foods, additives, and prepared meals. For others, however, it is the continuation of a way of life dictated by today's rural lifestyles, economic disadvantages, superstitions, and unavailability of modern medical facilities. Descriptions of the folk remedy and medicinal value of plants included in this book are found under "Comments" in the species descriptions.

POISONOUS PLANTS

Human poisoning by plants has been a serious health problem throughout history. In the United States thousands of people are poisoned annually. The majority of poisoning results from two situations: (1) small children grabbing brightly colored plant parts, such as berries, and eating them; and (2) wild food aficionados mistaking poisonous plants for edible ones. With the recent "return to nature" movement that has become a way of life for some people, both types of poisoning have been on the increase. But there is no need to refrain from increased use of the outdoors. Its therapeutic effect in times of stress in an increasingly urbanized world far outweighs the rare chance that accidental poisoning may occur. If one takes proper precautions, accidental poisoning need not occur.

The best deterrent to accidental poisoning is to become familiar with the poisonous plants. *Wildflowers of the Carolina Lowcountry* can be used as a guide to many of the poisonous plants that occur locally. Forty-nine of the 437 species included in this book are listed as poisonous, and their identification can be readily made from the photographs. Under the heading "Comments" in the species descriptions is a note on whether a species is poisonous. In addition, the bibliography lists several books that include a comprehensive account of the poisonous plants that occur in our area. When one considers, however, the many species of plants that are known to be poisonous worldwide (and undoubtedly others have not been identified), it is beyond the layperson to be familiar with all the poisonous plants they may come into contact with. Even experienced botanists are not fully aware of all the poisonous plants in their area. The following may serve as a precautionary guide:

1. Do not suck the nectar of flowers since it may be poisonous.
2. Never eat any part of an unknown plant.
3. Never chew or suck on jewelry made from imported fruits or seeds.
4. Never use twigs as skewers for cooking over open fires because it may poison the food.
5. Avoid smoke from campfires since it may carry toxic substances from the burning wood.
6. Never make or drink a tea made from an unknown plant.

It would be easy if we could look at a plant and determine from its physical features whether or not it is poisonous. Unfortunately, this is not possible, for nature has hidden many clues to the poisonous nature of plants. Consider the following:

1. The ripe fruit of the may-apple is edible, while the unripe fruit is poisonous.
2. The pulp of the apple is edible, while the seeds are poisonous.
3. The rhizome of Jack-in-the-pulpit is poisonous, but cooking the rhizome removes the poison.
4. The same amount of poison that may cause illness in a child may have no effect on an adult.
5. Plants that are bitter to taste may be harmless, while plants that are pleasant to taste may be poisonous.
6. Plants that are harmless to animals may be poisonous to people.
7. A substance may be relatively harmless if ingested after a heavy meal since it could be diluted by the stomach contents and be less readily absorbed, while on a empty stomach it could be readily absorbed and be poisonous.
8. The young leaves of some plants are harmless (and often eaten), while the older leaves may be poisonous (for example, pokeweed).
9. Not all plants with milky juice are poisonous. While the milky juice from the poinsettia causes a dermatitis in some people, the basis of chewing gum is the milky juice of a tree native to Mexico.
10. Three members of the genus *Rhus*—poison ivy, poison oak, and poison sumac—cause dermatitis, but another member—winged sumac—is harmless. In fact, its ripe berries are used to make a drink.

PART TWO

Guide to Species

Species Descriptions

THE SEASIDE GARDENS

1 *Coastal Beaches*

2 Sea Rocket

Cakile harperi Small
Brassicaceae (Mustard Family)
Flowers: March–October
Description: Smooth, fleshy, freely branched annual, rarely woody at base; up to 30" tall.
Range-Habitat: A coastal plain species; common along the coast in the coastal beach and adjacent coastal dune community.
Comments: Young cooked plants are of good quality but without a distinctive taste. The fleshy young foliage and young fruits are palatable when mixed as a salad with milder leaves; eaten raw they have the flavor of horseradish.

The fruit of sea rocket is indehiscent and divided into two segments. The terminal segment becomes dry and corky at maturity, breaking off from the basal segment and able to float great distances. The basal segment usually falls later and does not travel far.

Taxonomy: Two manuals, Godfrey and Wooten (1981) and Duncan and Duncan (1987), list this plant as a subspecies: *C. edentula* subsp. *harperi* (Small) Rodman.

3 Russian Thistle

Salsola kali L.
Chenopodiaceae (Goosefoot Family)
Flowers: June–Frost
Description: Herbaceous, freely branching annual, 10–25" tall; leaves fleshy, awn-shaped and sharp-pointed; the flowers vary in color, changing from whitish or gray to yellowish gray or pink in the fall.
Range-Habitat: A coastal plain species; common along the coastal beach.
Comments: Russian thistle is a native of Eurasia. Fragments of the plant that fall on the sand can be hazardous to exposed skin.

The stems turn from green to red or pinkish purple

in the fall. The dead plants are often blown loose and tumble down the beach in the process disseminating the seeds.

4 *Coastal Dunes*

5 **Devil-joint**

Opuntia drummondii Graham
Synonym: *Opuntia pussila* (Haw.) Haw.
Cactaceae (Cactus Family)
Description: Perennial, fleshy, leafless plant, creeping, often mat-forming; stems photosynthetic and segmented into loosely attached joints that separate readily; scattered over the stem are clusters of hairlike spines (glochids) with or without 2–4" long, sharp spines.
Range-Habitat: A coastal plain species; common on coastal dunes and sandy maritime forest areas.
Comments: Mechanical injury to people from devil-joint is common. The plant is often inconspicuous and hidden in the dune vegetation where it is stepped on with bare feet, inflicting a painful wound; or, due to the readiness with which the joints become detached, it may become embedded in shoes, after which the spines can work their way through if the shoes are thin.

Duncan and Duncan (1987) give a photograph of devil-joint in flower. Radford et al. (1968) state that the flowers "are not seen." I have never observed devil-joint in flower along the Carolina coast.

6 **Seaside Pennywort**

Hydrocotyle bonariensis Lam.
Apiaceae (Parsley Family)
Flowers: April–September
Description: Smooth, fleshy perennial rooting from the nodes of slender, creeping stems; leaves simple and peltate.
Range-Habitat: Outer coastal plain species; common on stable coastal dunes, swales, and moist, open, sandy areas.
Comments: The flowers and fruits of seaside pennywort are often present at the same time, and the compound umbel continues to produce new sections with new flowers.

7 **Beach Evening-primrose**

Oenothera drummondii Hooker
Onagraceae (Evening-primrose Family)
Flowers: April–October
Description: Spreading to creeping, densely hairy perennial, sometimes appearing shrubby in mild winters; flowers about 3" wide, turning toward the sun.
Range-Habitat: Infrequent along the coast but common on barrier islands. Also occurs in sandy, disturbed areas on developed barrier islands.
Similar Species: Dunes evening-primrose, *Oenothera humifusa*

Nuttall, is common on the dunes and has flowers about 1" wide, yellow tinged with pink; flowers May–October.

8 **Gaillardia; Fire-wheel**

Gaillardia pulchella Foug.
Asteraceae (Aster or Sunflower Family)
Flowers: April–Frost
Description: Short-lived, hairy, perennial herb, 6–28" tall; creeping to erect.
Range-Habitat: Common coastal plain species; in beach dunes, roadsides, and sandy habitats.
Comments: Apparently a southwestern species escaped from cultivation and naturalized in the above habitats (Justice and Bell, 1968; Rickett, 1967).

Many color variants and combinations of flowers occur in the red-pink-yellow range.

9 **Sand Ground-cherry**

Physalis viscosa subsp. *maritima* (M. A. Curtis) Waterfall
Solanaceae (Nightshade Family)
Flowers: May–September
Description: Rhizomatous perennial, 8–25" tall; star-shaped hairs throughout; fruit is a berry enclosed by a paperlike sac derived from the expanded, united sepals.
Range-Habitat: Outer coastal plain species; common on coastal dunes, roadsides, and sandy soils in maritime forests.
Comments: The ripe berries when cooked make a very palatable preserve. Otherwise, the plant is potentially toxic.

10 **Silver-leaf Croton**

Croton punctatus Jacquin
Euphorbiaceae (Spurge Family)
Flowers: Late May–November
Description: Annual or short-lived perennial, to 3' tall; entire plant (except upper leaf surface) covered with dense layer of small scales and glands that give the plant a brownish gray appearance; inconspicuous male and female flowers separate; fruit is a three-lobed capsule about .5" wide.
Range-Habitat: Restricted to the outer coastal plain; common on coastal dunes and beaches.
Comments: No other plant on the dunes even closely resembles this plant. Perennial plants can overwinter and produce new shoots from the old stems.

11 **Sea Purslane**

Sesuvium portulacastrum L.
Azioaceae (Carpetweed Family)
Flowers: May–Frost
Description: Fleshy, smooth, perennial herb with elongated, creeping branches rooting at the nodes and forming mats; leaves opposite; flowers and fruits on distinct stalks.
Range-Habitat: Outer coastal plain species; common in dune

swales, coastal dunes, high salt marshes, and beaches.
Comments: When growing in sandy sites, it forms mounds where sand, deposited by winds, builds up around the plant.
Similar Species: *S. maritimum* (Walter) BSP. grows in similar habitats. It differs in having flowers and fruits sessile and being an annual, erect to spreading, and not rooting at the nodes.

12 **Dune Spurge; Seaside Spurge**
Euphorbia polygonifolia L.
Synonym: *Chamaesyce polygonifolia* (L.) Small
Euphorbiaceae (Poinsettia Family)
Flowers: May–Frost
Description: Smooth, creeping to ascending annual with stems radiating from a single root; rare individuals may overwinter as a weak perennial; flowers are almost inconspicuous.
Range-Habitat: A coastal plain species; common on dunes.
Comments: Milky juice may cause skin irritation; plants, when eaten, may cause severe poisoning.
Similar Species: *E. ammannioides* HBK. often grows in association with dune spurge and has a similar growth pattern. The two species may be separated by seed and fruit size.

13 **Frog-fruits**
Lippia nodiflora (L.) Michaux
Synonym: *Phyla nodiflora* (L.) Greene
Verbenaceae (Vervain Family)
Flowers: Late May–Frost
Description: Perennial herb with creeping stems rooting at the nodes; flowers in dense heads on elevated stalks; leaves opposite.
Range-Habitat: Outer coastal plain species; common in sandy waste places, coastal dunes, dune swales, pond margins, and ditches.

14 **Spanish Bayonet**
Yucca aloifolia L.
Liliaceae (Lily Family)
Flowers: June–Early July
Description: Shrub (or tree) to 10' tall; stem coarse, usually with one to a few branches; leaves evergreen, dangerously sharp pointed; margins of leaves with sharp, spiny teeth that easily cut flesh; fruits maturing October–December.
Range-Habitat: An outer coastal plain species; common along the coast in active or stable dunes, Indian middens, abandoned house sites, and a variety of dry, inland habitats.
Comments: Spanish bayonet is naturalized from Mexico. The petals are used in salads or the entire flowers fried as fritters; the fruits can be cooked and eaten after the seeds are removed.

Similar Species: Two other species occur along the coast in similar habitats. *Y. gloriosa* L. is easily distinguished by having smooth leaf margins instead of having teeth; it blooms in October. Bear-grass (*Yucca filamentosa* L.), a native species, has its leaf margins frayed into filaments; it blooms in late April–early June. Both can be used as food like Spanish bayonet.

15 **Beach Pea**

Strophostyles helvola (L.) Ell.

Fabaceae (Bean Family)

Flowers: June–September

Description: Trailing, twining, or weakly climbing annual or perennial, herbaceous vine; leaves trifoliolate; petals rose to purple in color, often turning green.

Range-Habitat: A coastal plain species; common on the coastal dunes and open woods and clearings.

Comments: Pods and seeds can be cooked as vegetables.

16 **Common Marsh-pink**

Sabatia stellaris Pursh

Gentianaceae (Gentian Family)

Flowers: July–October

Description: Annual herb to 25" tall; flowers pink with yellowish, star-shaped center edged with red; stems freely branched with opposite leaves.

Range-Habitat: Outer coastal plain species; common in brackish swales within dune systems and brackish marshes.

17 **Sea Oats**

Uniola paniculata L.

Poaceae (Grass Family)

Flowers: June–November

Description: Coarse, rhizomatous perennial with stems 3–6' tall; rhizomes readily root at the nodes as the stem becomes covered with sand; reproduction is mainly by rhizomes.

Range-Habitat: A coastal plain species; common in coastal dunes and adjacent beaches.

Comments: Sea oats is tolerant of strong winds, sand abrasion, and saltspray and is one of the most important coastal plants in initiating dune formation and dune stabilization. In South Carolina it is unlawful to disturb sea oats on public property.

18 **Camphorweed**

Heterotheca subaxillaris (Lam.) Britton & Rusby

Asteraceae (Aster or Sunflower Family)

Flowers: July–October

Description: Polymorphic annual or short-lived perennial; glandular and sticky, with camphorlike odor when crushed;

stems to 6' tall, creeping-ascending or erect.
Range-Habitat: Common throughout the state; in the coastal area, common on coastal dunes and inland disturbed sites.
Comments: Duncan and Foote (1975) say the erect form occurs inland in disturbed sites (especially abundant in abandoned fields the first year after crops and in cutover woods), while the creeping-ascending form occurs in seaside habitats.

19 **Horseweed**
Erigeron canadensis var. *pusillus* (Nuttall) Ahles
Synonym: *Conyza canadensis* (L.) Cronq.
Asteraceae (Aster or Sunflower Family)
Flowers: July–Frost
Description: Bristly winter or summer, annual weed, 4–60" tall; basal leaves absent; stem leaves numerous, gradually reduced upward.
Range-Habitat: Common weed throughout the state; in the coastal plain it occurs in a variety of waste places and on coastal dunes and dune swales.
Taxonomy: Two varieties are recognized: var. *canadensis* with spreading pubescent stems and var. *pusillus* with smooth stems. Both varieties occur throughout the state; however, it is var. *pusillus* that grows in diminutive form on the coastal dunes and swales.
Comments: Horseweed is native to North America but has spread to Europe. Native Americans and early settlers used the plant to treat dysentery and sore throats. It may cause contact dermatitis.

Horseweed is an early successional plant in abandoned fields and becomes crowded out as perennials establish.

20 **Beach Morning-glory; Fiddle-leaf Morning-glory**
Ipomoea stolonifera (Cyrillo) Poiret
Convolvulaceae (Morning-glory Family)
Flowers: August–October
Description: Smooth, fleshy, trailing perennial, rooting at the nodes; most leaf blades lobed near the base, often deeply so.
Range-Habitat: A coastal plain species; infrequent in the coastal beach and adjacent coastal dune communities.

21 **Seashore-elder**
Iva imbricata Walter
Asteraceae (Aster or Sunflower Family)
Flowers: Late August–November
Description: Bushy-branched, perennial shrub to 3' tall; somewhat fleshy, smooth; commonly creeping at the base; tips of branches often dying during the winter.
Range-Habitat: A coastal plain species; common on coastal beaches, coastal dunes, and overwash areas.

22 **Dune Sandbur**
Cenchrus tribuloides L.
Poaceae (Grass Family)
Flowers: August–October
Description: Sprawling perennial or annual herb rooting at the

nodes; stem branches 4–28" long; the small flowers are enclosed within a spiny bur.

Range-Habitat: A coastal plain species; common on coastal dunes and adjacent beaches.

Comments: The spines can inflict painful puncture wounds on exposed skin and are equally painful to remove because of the backward-pointing barbs. The spines protect the plant from disturbance and provide an effective mechanism for seed dispersal. The burs turn from green to reddish with age and may remain on the dead stems throughout the winter.

23 **Seaside Panicum; Short Dune Grass**

Panicum amarum Ell.

Poaceae (Grass Family)

Flowers: September–October

Description: Rhizomatous perennial, usually rooting at lower nodes; stems usually solitary, 15–40" tall.

Range-Habitat: A coastal plain species; common in coastal dunes.

Comments: Seaside panicum becomes buried or uprooted when sand shifts extensively; therefore, it is not as effective as other dune grasses in binding soil.

24 **Sweet Grass**

Muhlenbergia filipes M. A. Curtis

Poaceae (Grass Family)

Flowers: October–November

Description: Tufted perennial to 40" tall; inflorescence a loose, limber panicle that turns pinkish when mature in the fall.

Range-Habitat: Outer coastal plain species; common along the coast in flats between coastal dunes, salt shrub thickets, and on stable dunes.

Comments: This is the famous sweet grass used by the basket makers in Mount Pleasant, South Carolina, to make the sweet grass baskets. Sweet grass is used in combination with longleaf pine needles, black needle rush, and cabbage palmetto leaves. There is concern that there will not be an adequate supply of sweet grass in the future to supply the basket makers. Inventories are being conducted along the coast to assess the natural populations, and transplanting experiments are testing the feasibility of establishing a cultivated source.

25 *Maritime Shrub Thickets*

26 **Wax Myrtle**

Myrica cerifera L.

Myricaceae (Wax Myrtle Family)

Fruits: August–October

Description: Aromatic shrub or small tree to 25' tall; male and female flowers on separate plants; leaves evergreen, although dropping in severe winters; leaves coated with orange, resinous glands on both surfaces.

Range-Habitat: A coastal plain species; common throughout a wide variety of habitats including maritime shrub thickets, maritime forests, salt shrub thickets, Indian middens, pine–mixed hardwood forests, loblolly pine plantations, and inland pine flatwoods.

Comments: Myrtle Beach, South Carolina, gets its name from this plant. The berries are covered with wax which in colonial days was used to make fragrant, greenish candles; the berries are still used to make candles. The wax may be irritating to some people.

The powdered root bark was an ingredient in "composition powder" once used as a folk remedy for chills and colds; the root bark was used to make an astringent tea and emetic.

27 **Dune Greenbrier**

Smilax auriculata Walter

Liliaceae (Lily Family)

Fruits: October–November

Description: Coarse-stemmed, evergreen vine; usually forming dense, low thickets; stems green, usually spineless; leaves smooth, green on both sides.

Range-Habitat: An outer coastal plain species; common in maritime shrub thickets, dunes, and maritime forests.

Comments: Its extensive rootstock helps it survive disturbance.

28 *Maritime Forests*

29 **Live Oak**

Quercus virginiana Miller

Fagaceae (Beech Family)

Description: Large- to medium-sized tree with wide, low-spreading branches and evergreen leaves.

Range-Habitat: Chiefly a coastal plain species; common in maritime forests, coastal hammocks, oak-hickory forests, along stream banks, and open, sandy woods.

Comments: The deep-grooved bark makes live oak a prime habitat for the establishment of epiphytes; occasionally Spanish moss, green-fly orchid, and resurrection fern grow on the same branch. The acorns, which ripen in the fall, are the sweetest of all the oaks and contain so little tannin that they can be eaten off the tree. The acorns are a major source of food for wildlife.

Live oak was an important lumber tree in colonial times. Curved pieces cut from the junction of the limb and trunk were used for ribs in wooden ships. Expeditions were sent from the North to the barrier islands of the Atlantic and Gulf Coast states to harvest live oak.

Contrary to popular belief, live oak is a fast-growing tree in rich soils with a normal life span of around 350 years. One only need observe the live oaks planted in the 1700s and 1800s along plantation avenues to understand how fast live oak grows. Widely planted as an ornamental, it withstands high winds, generally suffering only pruning. It also tolerates being inundated by salt water during hurricanes.

30 **Cabbage Palmetto**
Sabal palmetto Lodd. ex Schultes
Arecaceae (Palm Family)
Flowers: June–July
Description: Branchless tree to 65' tall; evergreen leaves fanlike at the top of the thick stems; leaf scars persist as shallow, incomplete rings on the trunk.
Range-Habitat: An outer coastal plain species; common in maritime forests on barrier islands, edges of ponds, and salt and brackish marshes.
Comments: Cabbage palmetto is a wind-adapted species; its soft "wood" allows it to bend with hurricane force winds and not be uprooted or broken. The trunks are used in the construction of wharves because it is not subject to injury from sea-worms. The inner portion of the apical meristem is very tender and palatable, resembling artichoke and cabbage in taste—hence, its common name. Unfortunately, removing the meristem kills the tree.

Cabbage palmetto is the state tree of South Carolina. During the Revolutionary War coastal forts were made of palmetto logs. The soft stems would absorb the force of cannon balls and not shatter. It is also prized as an ornamental tree.

31 **Hercules'-club; Toothache-tree; Pilentary Tree**
Zanthoxylum clava-herculis L.
Rutaceae (Rue Family)
Description: Shrub or small tree 20–30' tall; leaves alternate, odd-pinnately compound with thorny rachis; trunk of tree covered with pyramid-shaped, corky, spine-tipped outgrowths; flowers appear after the new leaves; fruit bears one black seed that hangs outside at maturity.
Range-Habitat: An outer coastal plain species; occasional in maritime forests, Indian middens, dunes, and sandy, thin woods.
Comments: The tree has a prominent place in American folklore. Francis Porcher (1869) lists numerous medicinal uses of this tree. Another source states that an oil derived from leaves and bark was used as a drug to treat toothache—hence, its common name. Chewing the leaves at first gives a pleasant sensation (makes the "tooth-fairies dance on one's tongue"), but then turns to a numbing sensation. The wood of this species has no commercial value.

32 **Resurrection Fern**
Polypodium polypodioides (L.) Watt
Polypodiaceae (Fern Family)

Description: Evergreen, epiphytic fern; leafstalks and underside of leaves covered with copious scales; rhizome creeping, scaly.

Range-Habitat: Common throughout the state on rocks, limbs, and crotches of large trees; in the coastal plain in maritime forests and any habitat with large, hardwood trees having a deep-grooved bark.

Comments: The common name comes from the curling of its leaves during prolonged drought with the lower side outward, giving it a dead-looking appearance. After a few hours of rain, the leaf absorbs water, uncurls, and is as alive and green as ever. Evidence was found in one study that suggested that the scales on the underside of the leaf hasten the recovery process by being channels of supplementary water absorption.

Resurrection fern spreads from tree to tree by wind-borne spores; once established on a limb, it spreads by its creeping rhizome.

33 **Spanish Moss**

Tillandsia usneoides L.

Bromeliaceae (Pineapple Family)

Flowers: April–June

Description: Rootless epiphyte on trees; stems usually curled, wiry; leaves filiform; leaves and stems bear gray scales that absorb atmospheric moisture and minerals; reproduction mainly vegetative by animals breaking and distributing strands of the plant; reproduction by seeds does occur; spent capsules remain throughout the winter.

Range-Habitat: A coastal plain species; common in maritime forests, swamp forests, and upland forests.

Comments: The cultural and economic uses of Spanish moss are legend. Its durable fiber is resistant to insects and highly resilient—characteristics which made it highly sought after as stuffing for mattresses and upholstering. It was used as a binder in construction of mud and clay chimneys.

Spanish moss is the most conspicuous epiphyte in the Lowcountry, and much of the charm and aesthetic appeal of the Lowcountry comes from the moss-draped live oaks so common along the plantation avenues.

34 **Wild Olive; Devilwood**

Osmanthus americana (L.) Gray

Oleaceae (Olive Family)

Flowers: April–May

Description: Small evergreen tree, occasionally reaching 30–40' in height; leaves opposite, simple; fruit a drupe, blue.

Range-Habitat: Chiefly an outer coastal plain species; common in maritime forests and dry woods.

Comments: The wood has no commercial value because of the small size of the tree. The wood is "devilishly" hard to split—hence, its common name. The fruits are eaten by birds and small mammals. Devilwood is occasionally used as an ornamental.

Manuals indicate the fruits mature in the fall, but there is confusion on this point. Coker and Totten (1945) report flowers and ripe fruits at the same time in April. I have observed trees with both ripe fruits and flowers in April. Apparently these were fruits from the preceding year that remained on the tree through the winter, a situation not uncommon if the fruits are not especially palatable to birds or other animals.

35 **Carolina Laurel Cherry**

Prunus caroliniana (Miller) Aiton

Rosaceae (Rose Family)

Flowers: March–April

Fruits: September–Spring

Description: Fast growing, short-lived evergreen tree to 40' tall; leaves alternate, entire, or with sharp teeth along the margin; black fruits (a drupe with thin, nearly juiceless flesh) stay on the tree until next flowering season.

Range-Habitat: Chiefly a coastal plain species; common in maritime forests, fence rows, thickets, and stable dunes.

Comments: It is probably native to the immediate coastal area but has been spread inland by birds from native populations and from cultivated plants. Laurel cherry is often planted as an ornamental and as a evergreen screen. The fruits are slightly poisonous to children but can be eaten with no ill effects by birds. Injured and wilted leaves are poisonous because they contain hydrocyanic acid that reacts with stomach acids to produce cyanide. Browsing animals have been killed by eating the mature leaves; the young leaves, however, can be safely eaten.

36 **Trailing Bluet**

Houstonia procumbens (J. F. Gmelin) Standley

Synonym: *Hedyotis procumbens* (Walter ex J. F. Gmelin) Fosb.

Rubiaceae (Madder Family)

Flowers: March–April

Description: Prostrate or creeping, perennial herb; flowers solitary on erect stalks; leaves opposite.

Range-Habitat: An outer coastal plain species; common in open, sandy sites in maritime forests and beach dunes.

37 **Bull Bay; Southern Magnolia**

Magnolia grandiflora L.

Magnoliaceae (Magnolia Family)

Flowers: May–June

Description: Large, fast-growing tree that produces flowers as early as ten years; leaves evergreen, persistent two years, shiny green above, often covered with reddish rust-covered hairs on the lower surface; flowers fragrant.

Range-Habitat: Chiefly a coastal plain species; common in

maritime forests, alluvial swamp forests, and beech forests.
Comments: Bull bay is planted extensively as an ornamental tree. Its leaves are used in wreaths and its flowers used for ornamental purposes. It is not abundant enough in any one site to be an important lumber source; the wood turns brown after exposure to air.

38 **Coral Bean; Cherokee Bean**
Erythrina herbacea L.
Fabaceae (Pea or Bean Family)
Flowers: May–July
Description: Perennial herb in the Lowcountry, 2–5' tall; flowers appear before the leaves; branchlets usually prickly; leaves alternate, trifoliolate; fruit pod constricted between the seeds and upon breaking open, brilliantly scarlet seeds often hang from the pod.
Range-Habitat: An outer coastal plain species; common in maritime forests, open dunes, and inland open, sandy woods; often persisting around abandoned house sites.
Comments: Coral bean is often cultivated in gardens. The seeds and bark possess alkaloids having curare-like action and may cause death if taken internally. The crushed stems are sometimes employed as fish poisons. In Mexico the seeds are used for poisoning rats and dogs. Coral bean is a woody shrub in Florida.

39 **Climbing Butterfly-pea**
Centrosema virginianum (L.) Bentham
Fabaceae (Pea or Bean Family)
Flowers: June–October
Description: Twining or trailing perennial, herbaceous vine from a tough, elongate root; leaves trifoliolate.
Range-Habitat: Common throughout the state; occurs in the coastal plain in oak-hickory forests, longleaf pine flatwoods, sandy and dry open woods, maritime forests, and coastal dunes.

40 **Beauty-berry**
Callicarpa americana L.
Verbenaceae (Vervain Family)
Fruits: August–October
Description: Deciduous shrub up to 8' tall; stems arching, with star-shaped hairs; leaves opposite, simple; flowers June and July, but are seldom noticed.
Range-Habitat: Common throughout the state; in the outer coastal plain it occurs in maritime forests, fence rows, barnyards, pine–mixed hardwoods, and oak-hickory forests.
Comments: Some references indicate the berries are edible but not very wholesome. Although sweet at first, they taste pungent and astringent afterwards.

41 *Salt Shrub Thickets*

42 Sea Ox-eye

Borrichia frutescens (L.) DC.
Asteraceae (Aster or Sunflower Family)
Flowers: May–September
Description: Rhizomatous shrub forming extensive colonies; little branched, 6" to 4' tall; leaves opposite, thick, somewhat fleshy; receptacle bracts hard and rigid, with sharp spine tips remaining on plant throughout the winter.
Range-Habitat: An outer coastal plain species; common in salt shrub thickets, edges of Indian middens and high salt marshes.
Comments: Along with groundsel-tree and marsh elder, sea ox-eye forms the highest zone in the salt shrub thicket. Of the three, this species occupies the lowest fringe next to the marsh.

43 Sand-vine

Cynanchium palustre (Pursh) Heller
Synonym: *Cynanchium angustifolium* Pers.
Asclepiadaceae (Milkweed Family)
Flowers: June–July
Description: Perennial, twining herb with opposite, linear leaves.
Range-Habitat: An outer coastal plain species; common in salt shrub thickets, Indian middens, and coastal hummocks; often overlooked.

44 Seaside Goldenrod

Solidago sempervirens L.
Asteraceae (Aster or Sunflower Family)
Flowers: Late August–November
Description: Fleshy-leaved, evergreen perennial; rhizomes short and stocky; 1–6' tall; leaves mainly basal, rapidly reduced upward.
Range-Habitat: An outer coastal plain species; common in salt shrub thickets, Indian middens, swales, and overwash areas.

45 Marsh-elder

Iva frutescens L.
Asteraceae (Aster or Sunflower Family)
Flowers: Late August–November
Description: Perennial, somewhat fleshy, bushy-branched shrub, 3–6' tall; leaves opposite; extremities dying during the winter.
Range-Habitat: An outer coastal plain species; common in salt shrub thickets, ditches in the salt marsh, and edge of Indian middens.
Comments: Along with sea ox-eye and groundsel-tree, marsh-elder forms the highest zone of the salt shrub thicket.

46 Groundsel-tree; Sea Myrtle; Consumption Weed

Baccharis halimifolia L.
Asteraceae (Aster or Sunflower Family)

Flowers: September–October
Description: Freely branched shrub 3–9' tall; leaves alternate, fleshy; larger leaves with few to several teeth; tardily deciduous, with some leaves hanging on throughout the winter; female plants have a satiny, white look in the fall from the mass of achenes tipped with bristles; male plants have a dull yellow appearance.
Range-Habitat: Throughout the coastal plain and piedmont; common in the outer coastal plain in salt shrub thickets, swales, fence rows, old fields, and pond margins.
Comments: Francis Porcher (1869) gives numerous medicinal uses of sea myrtle and states: "This plant is of undoubted valve, and of very general use in popular practice in South Carolina, as a palliative and demulcent in consumption and cough." One of its common names is based on this use.

In the fall during windy days in the Lowcountry, it is not unusual for the air to be filled with the achenes of sea myrtle.

It is thought that sea myrtle was originally a coastal plant but has spread throughout the state as disturbed areas increase.

47 **Yaupon**
Ilex vomitoria Aiton
Aquifoliaceae (Holly Family)
Fruits: October–November
Description: Evergreen shrub or small tree to 25' tall; male and female flowers on separate plants; leaves leathery, shiny above; fruit often persisting throughout the winter.
Range-Habitat: Chiefly an outer coastal plain species, but occurs in the sandhills region where it is rare; common in salt shrub thickets, maritime forests, Indian middens, maritime shrub thickets, and swamps.
Comments: The specific epithet, *vomitoria,* refers to the supposed emetic effect. Indians used a decoction of the dried old leaves, which were boiled down until the tea was very black and strong, to induce vomiting in purification rites. However, Hudson (1979) indicates that the emetic effect may have been the result of other herbs that were added to the drink. The young dried leaves have been and are still used today for a tea. The leaves are known to contain a considerable amount of caffeine, providing the lift people expect from tea. There were two attempts to grow yaupon commercially in Mount Pleasant in the early 1900s; both failed due to competition from oriental teas. But, yaupon does well in cultivation.

48 *Salt Marshes*

48 **Smooth Cordgrass**
Spartina alterniflora Loisel.
Poaceae (Grass Family)
Flowers: August–October
Description: Rhizomatous perennial; reproduction primarily by

spreading rhizomes.

Range-Habitat: Common marsh grass of the salt and brackish marshes of the coastal plain.

Comments: Smooth cordgrass is the dominant salt marsh plant, covering vast areas—most often to the exclusion of other species. Plants in the high salt marshes, especially along the edges of salt flats, may grow only a foot tall (the short form), while along the edges of tidal creeks it may grow 8' tall (the tall form).

Five major ecological roles have been assigned to salt marshes, of which smooth cordgrass plays the dominant role: formation of detritus, habitat for numerous animals, stabilization of coastal substratum through spreading rhizomes, filtration of coastal runoff, and removal of organic waste.

49 **Sea Lavender**

Limonium carolinianum (Walter) Britton

Plumbaginaceae (Leadwort Family)

Flowers: August–October

Description: Perennial herb, 6" to 2' tall, with basal rosette of leaves 2–10" long; flowers about 1/8" wide with white sepals and lavender to purple petals.

Range-Habitat: Common coastal plain species; grows in the short *Spartina* marsh, edges of salt flats, and edges of salt marsh thickets.

Comments: Salt glands on leaves and stems allow sea lavender to withstand being submerged in salt water. It varies in size and vigor depending on the habitat; about 6–8" high along the salt flats and up to 2' high on the edge of the salt shrub thicket.

50 **Saltmarsh Aster**

Aster tenuifolius L.

Asteraceae (Aster or Sunflower Family)

Flowers: Late August–November

Description: Perennial herb, 1–2' tall from slender, creeping rhizomes with curved, slightly zigzag spreading branches.

Range-Habitat: Common coastal plain species in the high salt marshes, salt shrub thickets, and dredged soil disposal sites.

Comments: Saltmarsh aster is never a major ecological component of the salt marsh. It is generally found in the high marsh dominated by smooth cordgrass and becomes conspicuous only when it blooms. In dredged soil disposal sites, it grows more robust and forms extensive stands.

51 *Salt Flats*

52 **Perennial Glasswort**

Salicornia virginica L.

Synonym: *Sarcocornia perennis* (Miller) A. J. Scott

Chenopodiaceae (Goosefoot Family)

Description: Fleshy, smooth halophyte; perennial, somewhat woody; stems trailing or weakly arching to erect, forming mats; present year's stem green, previous year's stem tan; leaf blade reduced to scales; flowers inconspicuous.
Range-Habitat: An outer coastal plain species; common in salt flats and high salt marshes.
Comments: Glasswort stems are filled with brine and make a pleasant, salty salad. It has also been popular as a source of pickles by first boiling the stems in their own salted-water before adding spiced oil or vinegar.
Along with saltwort, perennial glasswort is one of the few plants that can tolerate the high salinity of salt flats.

53 **Saltwort**
Batis maritima L.
Bataceae (Saltwort Family)
Description: Fleshy halophyte; perennial shrub; stems trailing and rooting at the nodes, forming dense patches from which arise erect, flowering branches; leaves opposite.
Range-Habitat: An outer coastal plain species; common on salt flats and high salt marshes.
Comments: Along with glasswort, one of the few species that can tolerate the high salinity of salt flats.

54 *Shell Mounds*

55 **Shell-mound Buckthorn**
Sageretia minutiflora (Michaux) Trel.
Rhamnaceae (Buckthorn Family)
Flowers: August–October
Description: Sprawling, weak-stemmed shrub to 10' tall, with many short, thornlike branches; leaves opposite or nearly so; flowers very fragrant; fruit drupelike, purplish black, maturing in the spring.
Range-Habitat: An outer coastal plain species; occasional along the coast on Indian shell middens, shell banks, and calcareous bluffs and hammocks.
Comments: Shell-mound buckthorn is probably more abundant than indicated by manuals but is certainly not common. I have observed it on numerous sites along the coast, especially on middens. It has the habit of draping over other vegetation on the middens but is not a woody vine. The population on the Sewee Indian Midden in the Francis Marion National Forest appears not to have suffered ill effects by Hurricane Hugo which inundated the midden in September of 1989. Although many trees on the midden were uprooted, weak stems apparently gave buckthorn the flexibility to survive the wind and water surge.

56 **Southern Red Cedar**
Juniperus silicicola (Small) Bailey
Cupressaceae (Cypress Family)
Fruits: October–November

Description: Aromatic, evergreen tree, 40–60' high; leaves on seedlings spreading from the twigs giving the seedling a prickly feeling; leaves of mature tree are like short scales in close, overlapping pairs; often both juvenile and mature leaves occur on the same plant; mature female cones bluish black; berrylike male cones borne on separate trees.

Range-Habitat: An outer coastal plain species; common on shell mounds, maritime forests, hammocks, dunes, and salt shrub thickets.

Taxonomy: Two species of cedars occur in the coastal plain: southern red cedar and eastern red cedar (*J. virginiana* L.), the latter more in the inner coastal plain. The differences between the two do not seem to be clear, and some authors consider southern red cedar a variety of red cedar.

Comments: The wood of both species has an essential oil that makes it durable—that is, not readily attacked by fungi or insects; hence, cedar has been used for shingles, fence posts, and cedar chests. Until the larger trees with straight trunks were exhausted, its wood was used extensively for making lead pencils. Cedar was also used by early Charleston furniture makers until mahogany became available. Today it is more of a specialty wood used for cedar chests and interior trim. The mature cones are eaten by many kinds of mammals and birds, including the cedar waxwing. Cedar is now grown on plantations for Christmas trees.

THE RIVER GARDENS

57 *Brackish Marshes*

58 Southern Wild Rice; Water Millet

Zizaniopsis miliacea (Michaux) Doell & Ascherson
Poaceae (Grass Family)
Flowers: May–July

Description: Coarse perennial from scaly rhizomes; commonly forming large, dense colonies; stems to 10' tall; on each branch, male flowers below and female flowers above.

Range-Habitat: Chiefly a coastal plain species; common in brackish and freshwater tidal marshes.

59 Wild Rice; Indian Rice

Zizania aquatica L.
Poaceae (Grass Family)
Flowers: May–October

Description: Coarse and robust perennial to 10' tall; often creeping and rooting at the nodes; lower branches of panicle widely spreading, upper branches ascending; male flowers hang from the lower branches of the inflorescence, while the female flowers are erect on the upper branches.

Range-Habitat: An outer coastal plain species; common in brackish marshes and tidal freshwater marshes.

Comments: Wild rice is an important food crop for both animals and people. American Indians used the grain to thicken soup, make bread flour and to cook game; today it is marketed as wild rice. Its main area of distribution is the Great Lakes and upper Mississippi region. The fruits are ready for harvest in mid-summer and fall. There is not a great enough quantity to make wild harvest feasible in our area. Attempts have mostly been futile in trying to cultivate the crop for commercial harvest.

Wild rice is not abandoned rice plants from the rice-growing era in South Carolina. The commercial plant of that time was *Oryza sativa* of Far Eastern origin which does not persist after cultivation.

60 **Needle Rush; Black Needle Rush**

Juncus roemerianus Scheele
Juncaceae (Rush Family)
Flowers: May–October
Description: Rigid and coarse perennial plant, from long, scaly rhizomes; stems grayish green to 5' tall; leaves round, stiff, and pungent, with sharp pointed tip.
Range-Habitat: An outer coastal plain species; common in brackish marshes and high salt marshes; often forming solid, vast stands.
Comments: Needle rush is characterized by having a hard, sharp point on the tip of the leaves. A person leaning over in the marsh could easily have an eye punctured.

61 **Soft-stem Bulrush; Great Bulrush**

Scirpus validus Vahl
Cyperaceae (Sedge Family)
Flowers: June–September
Description: Coarse rhizomatous perennial; stems round, spongy inside to 10' tall; leaves reduced to membranous sheaths.
Range-Habitat: Scattered throughout the state but more common in the outer coastal plain; common in brackish marshes, streams, and ponds; most often in shallow water.
Comments: The rootstock of great bulrush was used by the Indians; it was dried into a powder to make bread flour. The bruised young roots, boiled in water, made a sweet syrup.

62 **Leafy Three-square; Salt-marsh Bulrush**

Scirpus robustus Pursh
Cyperaceae (Sedge Family)
Flowers: July–September
Description: Rhizomatous perennial from thick rhizomes; 2–4' tall; stem sharply three-angled.
Range-Habitat: An outer coastal plain species; common in brackish marshes and high salt marshes.
Comments: This sedge is an extremely important food for wildlife. The seeds are food for ducks and marsh birds, and the stems and rootstocks are eaten by a variety of mammals and birds. On many plantations, brackish impoundments used to attract waterfowl are managed for reproduction of this species.

63 **Arrow-leaf Morning-glory**
Ipomoea sagittata Cav.
Convolvulaceae (Morning-glory Family)
Flowers: July–September
Description: Trailing or twining perennial vine; leaves arrow-shaped.
Range-Habitat: An outer coastal plain species; common in brackish marshes, dune swales, and the edge of brackish ponds.

64 **Saw-grass**
Cladium jamaicense Crantz
Cyperaceae (Sedge Family)
Flowers: July–October
Description: Leafy-stemmed perennial from large stolons; plant up to 10' tall, in dense tufts; leaves with margins and midribs with dangerous saw-teeth.
Range-Habitat: An outer coastal plain species; common in the brackish marshes along the coastal rivers; also a minor component in the tidal freshwater marshes.
Comments: This is the principal plant of the Everglades of Florida. When the Everglades are spoken of as the "river of grass," it is a misnomer because saw-grass is not a member of the grass family, but instead is a member of the sedge family.

65 *Tidal Freshwater Marshes*

66 **Native Wisteria**
Wisteria frutescens (L.) Poiret
Fabaceae (Bean Family)
Flowers: April–May
Description: High-climbing, woody vine; leaves alternate, odd-pinnately compound; legume matures in June–September.
Range-Habitat: Chiefly a coastal plain species; common in outer coastal plain in tidal freshwater marshes and swamp forests.
Comments: This is a native species of *Wisteria.* The legumes of the two commonly cultivated species, Chinese wisteria (*W. sinensis*) and Japanese wisteria (*W. floribunda*), are considered poisonous; although there is no mention in the literature of the native species, it should also be considered poisonous.

67 **Indigo-bush**
Amorpha fruticosa L.
Fabaceae (Bean Family)
Flowers: April–June
Description: Deciduous shrub, 5–12' tall; leaves pinnately compound with eleven to twenty-seven leaflets; one petal (the standard) wraps around the ten orange stamens and the style; other petals absent.

Range-Habitat: Common throughout the state along stream and river banks and in freshwater tidal marshes.
Comments: Often cultivated in the northeast as an ornamental.

68 Alligator-weed

Alternanthera philoxeroides (Martius) Grisebach
Amaranthaceae (Amaranth Family)
Flowers: April–October
Description: Emersed, perennial, aquatic herb with creeping stems rooting at the nodes or free floating in mats; stems to 3' long and forming dense mats; one source states viable seeds have not been found in the United States; therefore, reproduction is by vegetative means.
Range-Habitat: Chiefly a coastal plain species; common in various freshwater habitats such as tidal freshwater marshes, ditches, ponds, and swamps,
Comments: Alligator-weed is native to South America and was introduced into the United States in the early 1950s. It grows in a wide range of water and soil conditions. Mats of the plant can quickly block canals and ditches thereby reducing water flow and boat movement. Alligator-weed is not used as food by wildlife.

69 Water-spider Orchid

Habenaria repens Nuttall
Synonym: *Platanthera repens* (Nuttall) Wood
Orchidaceae (Orchid Family)
Flowers: April–Frost
Description: Terrestrial or aquatic herb; slender or stout, leafy; .5–3' tall; lower stem often producing elongate stolons with plantlets forming at the tips, especially when in floating mats or in soft substratum.
Range-Habitat: Chiefly a coastal plain species; common in a variety of habitats: tidal freshwater marshes, ditches and canals, and muddy shores of lakes, ponds, and streams.
Comments: Water-spider orchid is often found unexpectedly while looking for some other plant since it is quite inconspicuous. In the tidal freshwater marshes it is very difficult to spot water-spider orchid because of the dense growth or emergent species.

An interesting feature of this orchid is its adaptation as an aquatic plant. In the reservoirs created for the inland rice culture or similar impoundments, water-spider orchid often grows on large floating mats of vegetation in association with other aquatic plants.

70 Spider-lily

Hymenocallis crassifolia Herbert
Amaryllidaceae (Amaryllis Family)
Flowers: Mid-May–June
Description: Bulbous, smooth, perennial herb to 2' tall; leaves linear, all basal; unusual arrangement of stamens in which

the lower portion of the filaments are united into a thin, membranous crown and the upper filaments extend beyond the crown.

Range-Habitat: This species is confined to the outer coastal plain. It is common in tidal freshwater marshes and infrequent in brackish marshes and swamp forests.

Taxonomy: There is much confusion in the literature in the treatment of the genus *Hymenocallis.* No attempt can be made here to present a summary. The treatment by Radford et al. (1968) gives the name *H. crassifolia* for the species growing in the outer coastal plain.

Comments: Spider-lily is one of the most spectacular of the river marsh plants; often it bends over the water from the bulbs embedded in the river bank. I have observed spider-lily in virtually every freshwater river system in the coastal area of South Carolina.

71 **Arrow Arum**

Peltandra virginica (L.) Kunth

Araceae (Arum Family)

Flowers: May–June

Description: Emergent, perennial, upright herb from a short, stout rootstock; inflorescence a spadix with male flowers above and female below; spathe green with pale to white wavy margins.

Range-Habitat: Arrow arum is common throughout the state; in the outer coastal plain it occurs in tidal freshwater marshes, and shallow waters of ponds, slow-moving rivers, and swamps.

Comments: The large berries containing numerous minute seeds are an important food for wood ducks and marsh birds. This herb often forms large colonies along shallow waterways.

72 **Common Cat-tail**

Typha latifolia L.

Typhaceae (Cat-tail Family)

Flowers: May–July

Description: Persistent, emergent, rhizomatous perennial up to 9' tall; small flowers in a dense, cylindrical spike with the lower portion containing the female flowers and the upper portion the male flowers.

Range-Habitat: Common throughout the entire state; in the coastal plain it occurs in tidal freshwater marshes, ditches, edge of ponds, river banks, and shallow waters of lakes.

Comments: All parts of the plant are edible at one time when properly prepared. The rootstock is mostly starch and in the past was ground into flour by the Indians; the young shoots can be eaten as a pot-herb, and the immature flower spikes can be fried as fritters. Indians also used the pollen to make breadstuffs. The early colonists also made use of the cat-tail for food.

Cat-tail can quickly form dense stands from its

creeping rootstocks in shallow water and provides habitat for marsh birds, especially the red-winged blackbird. Its use as a wildlife food is negligible. Dense stands may cause ecological problems by impeding water flow and crowding out other species.

73 **Swamp Rose**
Rosa palustris Marshall
Rosaceae (Rose Family)
Flowers: May–July
Description: Upright, rhizomatous shrub to 10' tall; prickles curved; fruit (a hip) is red, matures September–October.
Range-Habitat: Chiefly a coastal plain species in South Carolina; common in tidal freshwater marshes and along streams, ponds, and swamp forests.
Comments: The hip, like all rose hips, is rich in vitamin C and can be eaten, made into jams, or steeped to make rose hip tea.

74 **Obedient Plant; False Dragon-head**
Physostegia leptophylla Small
Synonym: *Dracocephalum leptophyllum* Small
Lamiaceae (Mint Family)
Flowers: Late May–July
Description: Perennial herb from slender rhizomes; stem erect, 3–4.5' tall; leaves opposite, thin and flexible, with blunt teeth tips; corolla bright lavender-pink to purple.
Range-Habitat: A coastal plain plant in wet muck or peat of tidal freshwater marshes, river swamps, and sloughs.
Comments: The treatment of obedient plant is taken from Godfrey and Wooten (1981). They state the taxonomy of *Physostegia* (or *Dracocephalum* by some authors) is difficult and varies throughout the literature. Its status in South Carolina is also confusing. Some sources consider it rare; however, I have found it in numerous river systems and suggest a status of infrequent to common.

The common name, obedient plant, comes from the fact that the flowers tend to stay in a new position for a while after they are twisted to one side. It is truly one of the most spectacular plants of the marsh.

75 **Water Hemlock**
Cicuta maculata L.
Apiaceae (Parsley Family)
Flowers: May–August
Description: Erect, branching perennial to 9' tall; leaves pinnately, bipinnately, or tripinnately compound; the plant can be readily identified by cutting lengthwise through the stem base and root to reveal its diaphragmed nature; stem is magenta-streaked, hollow.
Range-Habitat: Occurs throughout the state; in the outer coastal

plain, common in freshwater tidal marshes, swamps, stream banks, and low roadside ditches.

Comments: All parts of the plant contain cicutoxin, a poisonous compound; however, the roots are particularly potent as only a mouthful is sufficient to kill an adult. The roots can be mistaken for water parsnip (plate 82). Children making peashooters from the hollow stems have been poisoned. The plant is not related to true hemlocks (*Tsuga* ssp., tree species), but is related to poison hemlock (*Conium maculatum*)—the plant used to kill Socrates.

76 **Water Primrose**

Ludwigia uruguayensis (Camb.) Hara

Synonyms: *Jussiaea uruguayensis* Camb. and *Jussiaea michauxiana* Fernald

Onagraceae (Evening-primrose Family)

Flowers: May–September

Description: Perennial herb or partly woody plant; stems creeping or floating; floating leaves found in spring are ovate, but as the plant becomes more erect in the summer and produces the erect, flowering stems, the leaves elongate and become more elliptical; all vegetative parts of the plant are covered by straight, white hairs.

Range-Habitat: Rapidly spreading and common throughout the coastal plain; occurs in lakes, ponds, sluggish streams, and tidal freshwater marshes.

Comments: Introduced from South America, the plant can become established by its horizontal stems (or rhizomes) on the edge of a pond or stream bank. These stems can continue to grow into the open water where they intertwine and form floating mats. The mats become so dense that other plants become established. The resulting mat of vegetation changes the ecology of a lake by shading the water column and inhibiting the growth of flora and fauna below. The composition of organisms shifts to those that favor these new conditions.

In a few years time I have seen open creeks in tidal freshwater rice fields in the Cooper River become completely closed by mats of water primrose. This reduction of open water may significantly affect waterfowl. Boating can also be severely impaired.

77 **Pickerelweed**

Pontederia cordata L.

Pontederiaceae (Pickerelweed Family)

Flowers: May–October

Description: Emergent, soft-stemmed perennial from a thick, short rhizome; to 3' tall; one leaf not far below the inflorescence, the others basal; seeds mature late summer to early fall.

Range-Habitat: Occurs throughout the state; common in the outer coastal plain in a variety of aquatic habitats: tidal freshwater marshes, lakes, ponds, roadside ditches, and swamp forests.

Comments: The seeds of pickerelweed are a pleasant and hearty food; the young leaf stalks can be cooked as greens. The roots are inedible and produce a burning sensation if ingested.

Pickerelweed can be a serious weed in ditches because it blocks drainage. It can also be a problem in small ponds since it can cover the surface.

The common name comes from a fish called the pickerel, which often occupies the same habitat as pickerelweed.

78 **Wapato; Lance-leaved Sagittaria**

Sagittaria lancifolia L.

Alismataceae (Water-plantain Family)

Flowers: May–November

Description: Robust perennial with coarse, stout rhizomes; erect flowering stalk to 5' tall, overtopping the leaves; flowers in whorls with the lower whorls comprised of female flowers and the upper whorls male.

Range-Habitat: Common in the outer coastal plain in tidal freshwater marshes, ponds, lakes, ditches, and freshwater swamps.

Similar Species: A similar species, *S. graminea,* occurs in the same habitats; it is difficult to separate the two in the field. *S. lancifolia* has linear filaments and its erect flowering stalks reach 5' tall. *S. graminea* has filaments dilated at its base and is a smaller plant.

A third species, *S. montevidensis* (giant arrowhead), occurs in the same habitats. It is easily distinguished by its large, broadly ovate, arrow-shaped leaves.

79 **Jewelweed; Spotted Touch-me-not**

Impatiens capensis Meerb.

Balsaminaceae (Touch-me-not Family)

Flowers: May–Frost

Description: Fleshy annual to 6' tall with hollow stems; leaves alternate; one sepal forms a prominent sac at the base of flower which ends in a curled spur.

Range-Habitat: Common throughout the state; in the coastal plain in tidal freshwater marshes, stream margins, and alluvial swamps.

Comments: The first common name refers to the water-repelling nature of the leaves and stems. On horizontal surfaces, water often rolls into beads giving the appearance of jewels in reflected light. The second common name refers to the fruits which, when mature, elastically coil into five sections; when touched, the sections explode thereby expelling the seeds.

The stem juice has fungicidal properties and has been used to treat athlete's foot. The stem juice is also a well-known treatment for poison ivy.

80 **Ground-nut**
Apios americana Medicus
Fabaceae (Bean Family)
Flowers: June–August
Description: Perennial, twining, herbaceous vine, 3–10' long; roots with tuberous enlargements; leaves alternate, pinnately compound with five to seven leaflets.
Range-Habitat: Grows throughout the state; common in outer coastal plain in tidal freshwater marshes, swamp forests, and bottomland forests.
Comments: The root tubers are edible. Eastern Indians used them as a staple food source, and the Pilgrims relied on them during their first year in Massachusetts. The tubers very quickly became a food source of early European settlers. Eaten raw, they leave an unpleasant rubberlike coating in the mouth. When cooked, the tubers loose this coating. The tubers can be used in soups and stews or fried like potatoes, or they can be ground into flour and used for bread.

81 **Large Marsh-pink**
Sabatia dodecandra (L.) BSP.
Gentianaceae (Gentian Family)
Flowers: June–August
Description: Perennial herb to 3' tall, with slender, coarse rhizomes; flowers pink or rarely white.
Range-Habitat: Chiefly an outer coastal plain species; common in tidal freshwater marshes, ponds, roadside ditches, and longleaf pine savannas.

82 **Water Parsnip**
Sium suave Walter
Apiaceae (Parsley Family)
Flowers: June–August
Description: Rank, smooth, perennial herb, 2–6' tall with hollow, angled stems from fascicled, fibrous roots; emersed leaves pinnately compound, leaflets lanceolate; plants that grow in water may have submerged leaves cut into very fine segments.
Range-Habitat: Chiefly an outer coastal plain species in South Carolina; common in tidal freshwater marshes, muddy sloughs in swamps, pond edges, and creek margins.
Comments: The roots of water parsnip can be eaten as a vegetable; however, because of the resemblance to the poisonous water hemlock (*Cicuta maculata*, plate 75), positive identification is necessary.

83 **Swamp Rose Mallow; Wild Cotton**
Hibiscus moscheutos L.
Malvaceae (Mallow Family)
Flowers: June–September

Description: Robust, herbaceous perennial to 6' tall; petals white, less often pink, but always with a purple-reddish center.
Range-Habitat: Throughout the state; infrequent in the mountains; common in the piedmont and coastal plain; in the outer coastal plain it occurs in tidal freshwater marshes, edges of swamp forests, roadside ditches, and brackish marshes.
Taxonomy: Three subspecies are given by Radford et al. (1968). The description above is for subspecies *moscheutos*.

84 **Water Willow**
Decodon verticillatus (L.) Ell.
Lythraceae (Loosestrife Family)
Flowers: July–September
Description: Colonial, shrubby perennial with arching stems; stems spongy below water surface.
Range-Habitat: A coastal plain species common in tidal freshwater marshes, ponds, and shallow waters of open swamps.
Comments: Whenever an arching stem touches the water, air-filled spongy tissue may develop. This spongy tissue buoys the stem so that it may root and form a new arching stem. Despite the common name, water willow is not related to the true willows.

Water willow has a high light requirement and is not found in wooded swamps; it often occurs on floating mats of vegetation.

85 **Eryngo**
Eryngium aquaticum L.
Apiaceae (Parsley Family)
Flowers: July–September
Description: Slender to robust biennial (or maybe perennial) herb, 1–6' tall; flowers in heads.
Range-Habitat: Primarily an outer coastal plain species; common in tidal freshwater marshes, river banks, ditches, and ponds; also found in brackish marshes.
Comments: This is the only genus in the Apiaceae in which the flowers are not in umbels.

86 **Hairy Hydrolea**
Hydrolea quadrivalvis Walter
Hydrophyllaceae (Waterleaf Family)
Flowers: July–September
Description: Perennial, aquatic herb with fleshy, hairy, spiny stems ascending from a creeping base.
Range-Habitat: Chiefly in the coastal plain; scattered in the piedmont; occasional in tidal freshwater marshes, edges of ponds and lakes, and river bottoms.

87 **Marsh Bulrush**
Scirpus cyperinus (L.) Kunth
Cyperaceae (Sedge Family)

Flowers: July–September
Description: Coarse, persistent, emergent, perennial herb from short, thick rhizomes; 3–6' tall; tufted, often forming large colonies.
Range-Habitat: Common throughout the state; in the outer coastal plain it occurs in tidal freshwater marshes, ditches, and pond margins.
Comments: Marsh bulrush provides cover and a secondary food source for songbirds and waterfowl in wetland habitats.

88 **Seashore Mallow**
Kosteletskya virginica (L.) Presl ex Gray
Malvaceae (Mallow Family)
Flowers: July–October
Description: Perennial herb to 5' tall; each flower lasts only a day; petals range from pink, lavender to white.
Range-Habitat: An outer coastal plain species; common in freshwater tidal marshes and brackish marshes.

89 **Climbing Hempweed**
Mikania scandens (L.) Willd.
Asteraceae (Aster or Sunflower Family)
Flowers: July–October
Description: Perennial, herbaceous vine, twining clockwise; leaves opposite.
Range-Habitat: Common throughout the state, usually in wet habitats; in the outer coastal plain in freshwater tidal marshes, disturbed wooded habitats, brackish marshes, and cypress swamps.
Comments: Climbing hempweed often forms a dense growth over low vegetation.

90 **Cardinal Flower**
Lobelia cardinalis L.
Campanulaceae (Bluebell Family)
Flowers: July–October
Description: Erect, usually unbranched perennial from basal offshoots, not rhizomatous; 2–6' tall.
Range-Habitat: Cardinal flower is common throughout the state; in the outer coastal plain it occurs in freshwater tidal marshes, swamp forests, and stream banks.
Comments: Species of *Lobelia* indigenous to America were initially employed in medicines for various purposes; however, cases of death from overdoses of medicinal preparations were frequent. They are now best considered poisonous.

The common name alludes to the bright red robes worn by Roman Catholic cardinals.

Cardinal flower is pollinated by hummingbirds; most insects cannot penetrate the long, tubular flowers.

Cardinal flower is adaptable to cultivation and can be

used in a variety of locations. It can even survive in a pot with frequent watering.

91 **Fragrant Ladies' Tresses**

Spiranthes odorata (Nuttall) Lindley
Orchidaceae (Orchid Family)
Flowers: July–Frost
Description: Semi-aquatic to aquatic herb; erect, 2–3' tall; leaves borne on lower portion of stem, but often well up the stem when the plant grows in standing water; flowers emit a potent fragrance described by some authors as a mixture of vanilla, coumarin, and jasmine.
Range-Habitat: A coastal plain species; listed as occasional in most manuals, but I consider it common based on my own field experience; found in water and mud of swamp forests and tidal freshwater marshes.
Comments: Often this orchid is hidden in the tall emergent plants of the marsh; one must be circumspect when looking for it in this habitat.
Taxonomy: There is much confusion in the literature about the taxonomy of the genus *Spiranthes*. This taxon is given variety rank by some and listed as *Spiranthes cernua* var. *odorata*. For simplicity it is listed here as a separate species as described by Luer (1975).

92 **Dodder; Love Vine**

Cuscuta gronovii Willd.
Convolvulaceae (Morning-glory Family)
Flowers: August–October
Description: Hemiparasite; annual, twining, herbaceous vine; leaves reduced to a few minute scales; haustoria develop all along the stem where it is in contact with the host plant.
Range-Habitat: Scattered localities throughout the state; in the coastal plain it is parasitic on a variety of woody or herbaceous hosts in tidal freshwater marshes, wet ditches, brackish marshes, swamps, and pond margins.
Comments: The seeds of dodder germinate in the soil; later the seedling tip begins to undergo a spiraling movement which often brings it into contact with a suitable host. As soon as it wraps itself around the host and haustoria develop, the soil roots die and contact with the soil is lost. Water, minerals, and a limited amount of organic material is transferred through the haustorium from the host to the dodder.
Similar Species: Five species of dodder occur in the Lowcountry; however, all are difficult to distinguish from one another. All five species can readily be identified as dodders because of the orange to yellowish twining stems and clusters of small, white flowers.

93 **Giant Beard Grass; Plume Grass**

Erianthus giganteus (Walter) Muhl.
Poaceae (Grass Family)

Flowers: September–October
Description: Perennial, persistent, emergent grass to 10' tall; often forming clumps or large colonies.
Range-Habitat: Chiefly in the coastal plain and lower piedmont; common in a variety of habitats: drier sites in tidal freshwater tidal marshes, ditches, wet pine savannas, drainage canals, and moist to wet clearings.

94 **Climbing Aster**
Aster carolinianus Walter
Asteraceae (Aster or Sunflower Family)
Flowers: Late September–October
Description: Robust, branching, somewhat woody, sprawling perennial; stems 3–6' long.
Range-Habitat: Outer coastal plain species listed from Charleston, Berkeley, and Georgetown counties; although listed as rare, I have observed it in numerous river systems in tidal freshwater marshes; also found in river swamps.
Comments: Climbing aster often forms robust growth in abandoned tidal rice fields on weathered posts, docks, etc.

95 **Bur-marigold; Beggar Tick**
Bidens laevis (L.) BSP.
Asteraceae (Aster or Sunflower Family)
Flowers: Late September–November
Description: Perennial (or maybe annual) herb from rhizomes; stems ascending to 3' tall, often creeping at base and rooted at nodes; often forming dense colonies; rays appear lighter yellow at the tips.
Range-Habitat: Primarily an outer coastal plain species; common in tidal freshwater marshes and shallow ponds filled with marsh vegetation.

96 *Tidal Freshwater Swamp Forests*

For species, see the bottomland swamp forests and the tidal freshwater marshes.

97 *Stream Banks*

98 **River Birch**
Betula nigra L.
Betulaceae (Birch Family)
Description: Small- to medium-sized, deciduous tree, often 60–80' tall; bark of trunk peels off in thin, paperlike layers.
Range-Habitat: Throughout the state, but chiefly piedmont and coastal plain; common along stream banks; generally associated with wet areas of river systems.
Comments: River birch is often cultivated because of its attractive, peeling bark. The inner bark is edible and makes a good emergency trail food. American Indians made use of the inner bark by drying it, then grinding it into flour. A

refreshing sap can be drunk from the tree in the spring, or boiled down into a sweet syrup. A tea can be made from the inner bark by simply boiling it in water. The dry, outer bark makes a good fire-starter. The wood is too knotty to be used as lumber.

99 **Tag Alder; Hazel Alder**

Alnus serrulata (Aiton) Willd.
Betulaceae (Birch Family)
Flowers: February–March
Description: Deciduous shrub to 30' tall; leaves alternate; usually growing in clumps; flowers in elongate conelike spikes; male spikes conspicuous in the spring before leaves appear; female "cones" persist through the winter after shedding seeds.
Range-Habitat: Common throughout the state; in the coastal plain found along stream banks, freshwater marshes, and wet places in forests.

100 **Button-bush**

Cephalanthus occidentalis L.
Rubiaceae (Madder Family)
Flowers: June–August
Description: Deciduous shrub or small tree 25–30' tall; leaves opposite or whorled; flowers in dense, round heads.
Range-Habitat: Throughout the state, but infrequent in the mountains; in the coastal plain common along sandy stream banks and margins of lakes, creeks, and freshwater marshes.
Comments: Both Francis Porcher (1869) and Morton (1974) give a variety of folk remedies for button-bush. Currently in the Lowcountry a "root tea" is gargled and swallowed to relieve throat irritation and to bring up phlegm from lungs in cases of severe colds. Historically the root was boiled with honey and comfrey making a pleasant syrup to treat diseases of the lungs (consumption).

Ducks, especially mallards, feed on the seeds of button-bush. Where it grows as dense thickets, button-bush serves as cover and as a nesting site for birds.

101 **Water Ash**

Fraxinus caroliniana Miller
Oleaceae (Olive Family)
Fruits: July–October
Description: Small, deciduous tree; leaves opposite, pinnately compound with five to eleven leaflets; fruit (samara) winged to base.
Range-Habitat: Chiefly a coastal plain species; common on sandy stream banks and as a understory tree in alluvial swamp forests.
Comments: Although the genus *Fraxinus* has many species that are valuable lumber trees, water ash is too small to be of any commercial value.

102 **American Sycamore**
Platanus occidentalis L.
Platanaceae (Sycamore Family)
Fruits: October
Description: Large, deciduous tree to 115' tall; outer bark separating into large, thin scales that fall away and expose the lighter, inner bark; male and female flowers produced on the same plant in separate, dense heads in April and May; female cluster develops into a hard fruit ball (that breaks apart in the fall) composed of numerous packed, narrow fruits.
Range-Habitat: Common throughout the state; in the coastal plain along sandy stream banks and hardwood bottoms.
Comments: Sycamore is often cultivated as an ornamental for the peeling bark. The tree is of little value for wildlife and birds and has never been an important lumber tree.

THE FRESHWATER AQUATIC GARDENS

103 *Freshwater Aquatics*

104 **Duckmeat; Duckweed**
Spirodela polyrhiza (L.) Schleiden
Lemnaceae (Duckweed Family)
Description: Free-floating aquatic; fronds about .2" long and bear two to four roots; fronds usually in groups of two to five plants, rarely solitary; flowers not seen.
Range-Habitat: Primarily a coastal plain species and common in pools, ponds, swamps, ditches, and margins of sluggish streams.
Comments: *Spirodela* can cause problems in small ponds which may become covered with a mass of duckmeat that interferes with livestock drinking and may clog irrigation pumps.

S. polyrhiza produces starch-filled turions during adverse environmental conditions of temperature and drought. The turion sinks to the bottom until favorable conditions return, at which time the turion expels a small bubble of gas that carries it to the surface where it germinates rapidly.

104 **Bog-mat**
Wolffiella floridana (J. D. Smith) Thompson
Lemnaceae (Duckweed Family)
Description: Fronds floating near the surface of the water, submerged; fronds rarely solitary, usually two or more attached by short, basal stalks forming a starlike colony; fronds .3–.6" long; no roots; reproduction is mostly vegetative by budding, but occasionally a frond produces one male and one female flower.
Range-Habitat: Common throughout the coastal plain in slightly acidic and highly organic waters of ponds, roadside ditches,

streams, swamps, and marshes.

Comments: Bog-mat most often occurs intermixed with the other three genera of Lemnaceae.

105 **Floating-heart**

Nymphoides aquatica (Walter ex J. F. Gmelin) Kuntze

Gentianaceae (Gentian Family)

Flowers: Late April–September

Description: Perennial herb; free-floating and anchored from a thick rhizome; leaves green above, usually purple below; stem terminates in an umbel of flowers and one leaf with a short stalk.

Range-Habitat: Primarily a coastal plain species; frequent in outer coastal plain in fresh waters of lakes, ponds, sluggish streams, and swamps.

106 **Cow-lily; Spatterdock**

Nuphar luteum (L.) Sibthrop & Smith

Synonym: *Nuphar advena* (Aiton) Aiton f.

Nymphaeaceae (Water-lily Family)

Flowers: April–October

Description: Perennial herb from a large rhizome; leaves either floating, submerged, or emersed; leaf blades stand erect above the water and are submerged or floating only during early growth or periods of high water; leaves are highly variable and may be wide or rather narrow.

Range-Habitat: Primarily a coastal plain species; common in alluvial areas such as pond margins; also grows well in sluggish canals, swamps, and ponds.

Comments: The Indians used the seeds and rhizomes as food. The rhizomes were roasted or boiled, after which they could be easily peeled; the sweet interiors were then cut up for soups and stews.

The plant spreads so quickly that waterflow in a canal or small stream is seriously curtailed two to three years after establishment.

Taxonomy: Much variation exists in the literature about cow-lily. Some authors recognize four subspecies of *Nuphar* while others elevate the four subspecies to species level.

107 **Purple Bladderwort**

Utricularia purpurea Walter

Synonym: *Vesiculina purpurea* (Walter) Raf.

Lentibulariaceae (Bladderwort Family)

Flowers: May–September

Description: Carnivorous, free-floating, rootless aquatic; usually in bunches or mats with flowering stalks supported above the water surface by the mass of submerged, branching stems that often terminate in bladders; flowering stalks grow to 4" tall with one to three flowers.

Range-Habitat: Scattered throughout the coastal plain in a variety of aquatic habitats; probably more common than indicated by manuals.

Comments: This is the only species of bladderwort in the Lowcountry with purple flowers. The underwater bladders catch minute insects.

108 **Floating Bladderwort**

Utricularia inflata Walter var. *inflata*

Lentibulariaceae (Bladderwort Family)

Flowers: May–November

Description: Free-floating, carnivorous herb; upper leaves, consisting of an inflated stalk and rachis, are whorled, forming a flotation device which supports the flowering stalk; submerged "leaves" bear the bladders that trap aquatic animals, providing minerals.

Range-Habitat: A coastal plain species common in swamps, lakes, ponds, roadside ditches, and pools.

Comments: An interesting feature of this bladderwort is the development of the stalk and flotation device from the submerged part of the plant in spring. The flotation structures develop at the end of the immature stalk while underwater. As both grow, their buoyancy causes them to rise to the surface, by which time the upper part of the stalk is fairly well developed. The result is the flowers (on the end of the stalk) are elevated above the water so pollination can occur. To suddenly behold a pond covered with a mass of floating bladderworts where none existed the previous day is one of the wonders of nature. This plant survives drought conditions by producing drought-resistant tubers; when submerged after the pond fills, they generate a new plant.

Taxonomy: Some manuals list two varieties of *U. inflata:* the one described above and a second one, *U. inflata* var. *minor* (L.) Chapman. The latter is much smaller and given species designation by some authors and called *U. radiata* Small. This smaller species is also a coastal plain species but based on my observation is infrequent. The habitat of both is similar.

109 **Water Hyacinth**

Eichhornia crassipes (Martius) Solms

Pontederiaceae (Pickerelweed Family)

Flowers: June–September

Description: Free-floating, aquatic herb with a basal cluster of leaves; often may be rooted in mud as the water recedes and may persist for several months.

Range-Habitat: Occasional and scattered throughout the coastal plain in ponds, ditches, canals, and abandoned rice fields; its abundance diminishes in cooler areas.

Comments: Water hyacinth was naturalized into Florida in 1884 and, through prolific growth, has rapidly become a serious weed which clogs waterways in the frost-free coastal areas of the Southeast.

The inflated, bulbous leaf stalks consist of aerenchyma tissue that gives the plant great buoyancy. Reproduction is mainly by vegetative means as sections break off and are carried by currents or driven by wind.

110 **Sacred Bean; Water Chinquapin**
Nelumbo lutea (Willd.) Persoon
Nymphaeaceae (Water-lily Family)
Flowers: June–September
Description: Rhizomatous, perennial herb; leaves produced early in the season lie on the water surface, and as the leaf stalk grows during the summer, the leaves extend above the surface; flowers solitary on the flowering stalk; leaves and flower often 3–4' above water surface.
Range-Habitat: Infrequent and scattered throughout the coastal plain in muddy areas of ponds, sluggish streams, and lake margins.
Comments: Water chinquapin was a favorite Indian food. The tender, immature seeds were eaten raw or cooked; the ripe seeds were parched to loosen the containing shell, then eaten dry, baked, boiled, or ground to make bread. The tuberous enlargements of the rootstocks become filled with starch in the fall and make a tasty food when baked or boiled and seasoned.

The pistil develops into a funnel-shaped fruit about 4" in diameter. The apex contains several cavities with each cavity containing a single seed which shakes out when ripe.

111 **Fragrant Water-lily; Sweet Water-lily**
Nymphaea odorata Aiton
Nymphaeaceae (Water-lily Family)
Flowers: June–September
Description: Perennial herb; free-floating and anchored from a thick rhizome; distinctive for its sweet-scented, white, showy flowers; leaves purple beneath.
Range-Habitat: Common throughout the coastal plain in pools, ponds, and sluggish streams.
Comments: Morton (1974) indicated that a root infusion of water-lily was used by rural people of the Lowcountry as treatment for diarrhea; for itching of private parts, they bathed with and/or drank a decoction of the root. Water-lily is used as an ornamental in small ponds where it often develops dense stands that may cover the water and interfere with boating and fishing.

112 **Pondweed**
Potamogeton plucher Tuckerman
Potamogetonaceae (Pondweed Family)
Flowers: June–September
Description: Perennial herb; free-floating and anchored from slender rhizome; leaves of two kinds—submerged and floating; floating leaves elliptic or ovate; submerged leaves lanceolate.

Range-Habitat: Common throughout the coastal plain in pond cypress swamps, ponds, ditches, and sluggish streams.
Similar Species: *P. diversifolius* Raf., which occurs in the same habitat, is distinguished by having thin, narrowly linear, submerged leaves.

113 **Frog's-bit**
Limnobium spongea (Bosc.) Steudel
Hydrocharitaceae (Frog's-bit Family)
Flowers: June–September
Description: Perennial herb, generally free-floating in dense, floating mats or becoming rooted in mud as a pond or marsh dries up.
Range-Habitat: A common coastal plain species occurring in a variety of shallow, quiet-water habitats such as swamps, ponds, drainage ditches, lakes, and marshes.
Comments: Two forms of frog's-bit occur during its life cycle: vegetative, floating, and basal clusters of nearly kidney-shaped leaves, which in turn give rise to more robust plants consisting of clusters of ascending and erect leaves on stalks on which flowers and fruits develop.

Extensive growth usually creates problems in management of small ponds.

114 **Water-shield**
Brasenia schreberi Gmelin
Cabombaceae (Water-shield Family)
Flowers: June–October
Description: Perennial, floating-leaved, anchored herb, the leaves and flowers arising from a slender, creeping rootstock with considerable branching of the stem; stems, leaf stalks, and underside of the leaves coated with a gelatinous material; underside of the leaf is purple, the upper side bright green.
Range-Habitat: Common coastal plain species in freshwater ponds, swamps, lake edges, and sluggish streams.
Comments: Often grows in extensive stands that exclude other vegetation in small ponds.

THE PEATLAND GARDENS

115 *Pocosins*

116 **Pond Pine; Pocosin Pine**
Pinus serotina Michaux
Pinaceae (Pine Family)
Female Cones: All Year
Description: Medium-sized, evergreen tree; cones top-shaped or almost globe-shaped; foliage leaves mostly in threes, 6–8" long.
Range-Habitat: A coastal plain and lower piedmont species; common in pocosins and pond cypress savannas; less common in longleaf pine savannas.

Comments: Pond pine is a "serotinous" species, depending on fire for regeneration. Its cones remained closed until fire softens the resinous seal which allows the cone to dry out. Internal stresses then force open the scales, allowing the winged seeds to escape. Fire also removes the underbrush and presents open, sunny conditions ideal for germination and establishment of the seedlings. Pond pine also has another adaptation to lands frequented by fire. It sprouts new branches along the trunk from latent axillary buds. Although the crown may have been destroyed by the fire, the new shoots are tied to a old, healthy root system and quickly develop into new branches.

Pond pine is not an important lumber species because it has poor growth form and grows in wet areas not easily accessible for timbering.

117 **Highbush Blueberry**

Vaccinium corymbosum L.

Ericaceae (Heath Family)

Flowers: Late February–May

Description: A highly variable, deciduous shrub with one to several trunks from the base, 5–15' tall; flowers precede or develop concurrently with the new leaves; berries mature in June–August.

Range-Habitat: A very common coastal plain species growing in pocosins and adjacent longleaf pine savannas; less common in longleaf pine flatwoods.

Comments: This is the native species from which the blueberry of commerce was derived. The berries are also an important food for a variety of wildlife; twigs and foliage are browsed by deer and rabbits. Highbush blueberry responds to burning, after which it produces its best crop of berries.

118 **Leather-leaf**

Cassandra calyculata (L.) D. Don

Synonym: *Chamaedaphne calyculata* (L.) Moench

Ericaceae (Heath Family)

Flowers: March–April

Description: Low, rhizomatous shrub, 1–3' tall; flowers on an arching stem; leaves evergreen.

Range-Habitat: Leatherleaf is a coastal plain species known only in Berkeley, Horry, and Marion counties. It is infrequent and grows in bogs and pocosins. One good site is Lewis Ocean Bay Heritage Preserve in Horry County.

119 **Creeping Blueberry**

Vaccinium crassifolium Andrews

Ericaceae (Heath Family)

Flowers: April–May

Description: Trailing shrub, with upright branches, usually rooting at the nodes; leaves evergreen; berry black, matures June–July.

Range-Habitat: Primarily a coastal plain species growing in pocosin ecotones and adjacent longleaf pine flatwoods and savannas; common but often overlooked because it grows close to the ground, hidden in vegetation.
Comments: The berry is edible.

120 **Witch-alder**
Fothergilla gardenii Murray
Hamamelidaceae (Witch-hazel Family)
Flowers: April–May
Description: Colonial shrub 1–3' tall; leaves with starlike hairs.
Range-Habitat: Primarily a coastal plain species; infrequent in pocosin ecotones and longleaf pine savannas.

121 **Possum-haw**
Viburnum nudum L.
Caprifoliaceae (Honeysuckle Family)
Flowers: April–May
Description: Deciduous shrub or small tree, 15–20' tall; leaves opposite; blue-black fruits mature in August–October.
Range-Habitat: Primarily a coastal plain species, but does occur in the mountains; common on the edge of pocosins, in longleaf pine savannas, and edges of swamps.
Comments: The ripe fruits have a thin and rather dry, sweet pulp which is palatable when eaten raw.

122 **Sweet Pitcher-plant**
Sarracenia rubra Walter
Sarraceniaceae (Pitcher-plant Family)
Flowers: April–May
Description: Carnivorous, flowering herb; perennial from rhizomes; leaves modified into hollow tubes, 4–20" tall, as passive traps to catch insects; flowering stalk usually exceeding the leaves; petals maroon on outer surface, greenish on inner.
Range-Habitat: A coastal plain species; frequent along edge of pocosins, in seepage areas, and in longleaf pine savannas.
Taxonomy: At one time a variety of *S. rubra* was recognized that occurred in the mountains, *S. rubra* var. *jonesii*. This variety is now recognized as a distinct species, *S. jonesii* Wherry. *S. rubra* is now considered confined to the coastal plain.
Comments: Sweet pitcher-plant mainly catches ants; the opening is too small to admit larger insects. It appears to grow more robust in seepage areas along the edge of pocosins and in sphagnum openings in pocosins. In the pine savannas where it is also found, it is often diminutive.

123 **Frog's Britches; Hunter's Cup**
Sarracenia purpurea L.
Sarraceniaceae (Pitcher-plant Family)
Flowers: April–May

Description: Rhizomatous, perennial, evergreen, carnivorous herb with hollow leaves modified as trapping structures; flowering stalks 8–16" tall.

Range-Habitat: Primarily a coastal plain species found in the Lowcountry counties of Georgetown, Clarendon, Horry, and Marion; rare; favors sphagnum openings in pocosins where it grows more robust; also found in moist, longleaf pine savannas.

Comments: Frog's britches differs from other pitcher-plants in the Lowcountry by having leaves that lie horizontally but curve upward; the hood is erect and does not cover the mouth, and its inner surface bears many stiff hairs which point downward toward the mouth. The open mouth permits the pitcher to fill with rainwater where insects that fall in are drowned. It is believed that a wetting agent secreted by glands into the water denies the insect buoyancy so that it cannot fly off the water's surface.

124 **Lamb-kill; Sheep-kill**

Kalmia angustifolia var. *caroliniana* (Small) Fernald
Synonyms: *Kalmia carolina* Small and *Kalmia angustifolia* L.
Ericaceae (Heath Family)
Flowers: April–Early June

Description: Evergreen, rhizomatous shrub, 1–3' tall, forming sizable colonies; leaves in whorls of three, or occasionally opposite or alternate on some branches.

Range-Habitat: Primarily a coastal plain species found in Clarendon, Marion, and Horry counties on the edge of pocosins and in longleaf pine savannas. Rare, but good populations occur in Horry County in the Cartwheel Bay Heritage Preserve and the Vaughn Tract of the Little Pee Dee River Heritage Preserve and on the edge of Bennett's Bay in Clarendon County.

Comments: The common name comes from the fact that it is poisonous to livestock. The flowers are miniatures of the larger mountain laurel (*Kalmia latifolia*). The ten stamens, whose anthers are tucked into pockets of the corolla, pop out when touched by insects causing pollen to spray the insect—an adaptation that assists in cross-pollination.

125 **Fetterbush**

Lyonia lucida (Lam.) K. Koch
Ericaceae (Heath Family)
Flowers: April–Early June

Description: Rhizomatous, evergreen shrub to 8' tall; usually forming dense colonies, especially in pocosins; first-year twigs strongly angled; leaves with a distinct marginal vein.

Range-Habitat: Fetterbush is a common coastal plain species found in pocosins, longleaf pine and pond cypress savannas, and swamps.

Comments: Fetterbush is often the dominant shrub in pocosins, forming dense colonies that make penetration difficult. It sprouts vigorously from rhizomes after a fire, quickly becoming reestablished.

126 **Honeycup**

Zenobia pulverulenta (Bartram) Pollard
Ericaceae (Heath Family)
Flowers: April–June
Description: Rhizomatous, deciduous shrub to 6' tall; leaves of two forms: either green on both surfaces or green above and bluish white beneath; both forms occur in the same site.
Range-Habitat: A coastal plain species; in South Carolina found only in the northeast half of the Lowcountry; common in pocosins.
Comments: This is one of our most spectacular ericaceous shrubs when in full flower.

127 **Sweet Bay**

Magnolia virginica L.
Magnoliaceae (Magnolia Family)
Flowers: April–July
Description: Small shrub or tree 30–80' tall; often a bushy stump-sprout in burned or cut-over areas; leaves entire, evergreen, or semi-evergreen, persisting into winter with a few remaining until spring; leaves white beneath, easily noticeable from a distance.
Range-Habitat: A coastal plain species common throughout; found in pocosins, longleaf pine and pond cypress savannas, and swamp and bay forests.
Comments: Often cultivated because of its showy, fragrant flowers.

128 **Poison Sumac**

Rhus vernix L.
Synonym: *Toxicodendron vernix* (L.) Kuntze
Anacardiaceae (Cashew or Sumac Family)
Flowers: May–Early June
Description: Smooth, tall shrub or small tree, 20–30' tall; leaves deciduous, produced simultaneously with flowers; leaves alternate, odd-pinnately compound, with seven to thirteen leaflets, the leaf stalks and young twigs nearly always reddish.
Range-Habitat: Common coastal plain species growing in pocosins (more commonly on the edge) and adjacent, acidic swamps.
Comments: The oils in all parts of the plant can cause severe dermatitis in susceptible individuals similar to poison oak and poison ivy, which are in the same family. Smoke from burning leaves or twigs can carry the volatile oil. The whitish, waxy fruits ripen in early fall but may persist throughout the winter where they are fed upon by numerous birds.

129 **Rose Pogonia**

Pogonia ophioglossoides (L.) Ker
Orchidaceae (Orchid Family)
Flowers: May–June

Description: Slender orchid, 3–24" tall with a single green leaf about halfway up the stem; stem supports a single flower subtended by leaflike bract.

Range-Habitat: Rose pogonia is primarily a coastal plain orchid; it is infrequent in seepage bogs on the edge of pocosins, openings in pocosins, poorly drained roadside ditches, and longleaf pine savannas.

130 **Swamp Honeysuckle; Clammy Azalea**

Rhododendron viscosum (L.) Torrey

Synonym: *Azalea viscosa* L.

Ericaceae (Heath Family)

Flowers: May–July

Description: Deciduous shrub, 3–9' tall; corolla white or rarely pink and covered with reddish, sticky hairs; twigs hairy; flowers appear after the leaves.

Range-Habitat: Common throughout the state; in the coastal plain on the edge of pocosins and pond or stream margins.

Comments: Swamp honeysuckle is the state's latest flowering azalea.

131 **Titi; Leatherwood**

Cyrilla racemiflora L.

Cyrillaceae (Titi Family)

Flowers: May–July

Description: Tall shrub or small tree to 25' tall; flowers in racemes clustered near the end of the previous year's twig; leaves semi-evergreen, some falling throughout the winter, but a few remaining until new ones appear in the spring.

Range-Habitat: Common coastal plain species in pocosins, edges of swamps, in longleaf pine flatwoods, and savannas.

Comments: Titi is a good ornamental because of its attractive flowers and leaves which turn orange and scarlet in the fall. The trees are good honey plants because the abundant flowers produce large quantities of nectar.

132 **White Arum**

Peltandra sagittaefolia (Michaux) Morong

Araceae (Arum Family)

Flowers: June–August

Description: Perennial herb from a short, stout rootstock; inflorescence a spadix subtended by a flared, open, white spathe; flowers of two sexes, with the male flowers on the upper part of the spadix and the female flowers on the lower part.

Range-Habitat: Listed as rare by most manuals and floras but probably should be listed as infrequent. I have encountered white arum in numerous localities in Berkeley and Charleston counties; white arum grows in sphagnum openings in pocosins and under powerlines created through pocosins.

Comments: All parts of the plant contain crystals of calcium oxalate which causes irritation of the mucus membranes of the mouth and throat, possibly leading to asphyxiation if a great quantity is eaten. Boiling of the rootstock removes the crystals, making them edible.

133 **Loblolly Bay**

Gordonia lasianthus (L.) Ellis
Theaceae (Tea Family)
Flowers: July–September
Description: Evergreen shrub or small tree up to 75' tall in rich sites; leaves simple, alternate, shallowly toothed, and smooth beneath; fruit a capsule that matures in September–October.
Range-Habitat: A coastal plain species common in pocosins and bay forests.
Comments: Loblolly bay is sometimes used as an ornamental because of its showy flowers; however, it does not grow well under cultivation except in moist sites. It has little food value for wildlife and little or no commercial use. The leaves often have a ragged appearance because of being chewed by insects.

134 **Bamboo-vine**

Smilax laurifolia L.
Liliaceae (Lily Family)
Fruits: September–October
Description: Robust, evergreen, high-climbing vine; dead stems intermixed with living, forming impenetrable thickets; stems most often spiny.
Range-Habitat: Throughout the state, but primarily a coastal plain species; common in pocosins, less common in swamps.
Comments: Bamboo-vine along with fetter-bush are the two species primarily contributing to the impenetrable nature of pocosins.

THE BOTTOMLAND FOREST GARDENS

135 *Bald Cypress–Tupelo Gum Swamp Forests*

136 **Mistletoe**

Phoradendron serotinum (Raf.) M. C. Johnston
Loranthaceae (Mistletoe Family)
Description: Evergreen shrub; obligate hemiparasite on a variety of broadleaf, deciduous trees; flowers small, male and female on separate plants, maturing September–November and sporadically through the winter; white berries maturing November–January, persisting into spring.
Range-Habitat: Common in the coastal plain and piedmont, less frequent in the mountains; on branches of broadleaf, deciduous trees exposed to sun in swamp forests and other

forested areas.

Comments: There are numerous species of *Phoradendron* in North America; however, this is the only species in South Carolina. This is the common mistletoe used in Christmas holiday decorations.

Mistletoes cause enormous economic loss in many parts of the world. In North America it is a troublesome pest in walnut, pear, and pecan plantations. In these situations, injury results from broken branches (which allow invasion of insects and fungi) and the slowed growth of the trees due to loss of water and minerals. In South Carolina little economic loss results from mistletoe since it affects trees that are of little commercial value. (It does not attack pines.)

The berries are covered with a sticky material poisonous to man; poisoning often occurs during Christmas when the plant is used for decorations and children eat the berries. Its one-seeded berries, however, can be eaten by a wide variety of birds that use the pulp for food. The birds spread the seeds through their droppings and by wiping their beaks on branches. In both cases, germination occurs on the branch, the haustorium penetrates the host tissue, and the xylem bridge forms between the host and mistletoe.

137 **Bald Cypress**

Taxodium distichum (L.) Richard

Taxodiaceae (Taxodium Family)

Female Cones: All Year

Description: Deciduous conifer with two-ranked leaves (needles); large tree, 70–130' tall; separate male and female cones on the same tree; female cones green, turning brown at maturity in the fall.

Range-Habitat: Chiefly a coastal plain species; common throughout in alluvial and nonalluvial swamp forests and hardwood bottoms.

Comments: Bald cypress historically was one of the most important lumber trees of the South. Its wood is very durable due to the presence of essential oils and was used for shingles, barrels, caskets, and beams. Indians and early settlers carved cypress logs into boats, troughs, and washtubs. Cypress was also used to make the rice field trunks in the 1700s and 1800s, many of which still persist today along the coastal rivers. Much of the fine Charleston-made furniture has cypress as the secondary wood.

Cypress as a major commercial timber source has decreased over the years because the tree does not reproduce well after clear-cutting. Large stands of bald cypress exist today mainly on state and federal lands where it is protected.

When growing in water, the truck of bald cypress forms a basal buttress and its roots produce knees projecting above water. Bald cypress is also planted as an ornamental because it grows well on a variety of upland soils.

138 **Golden-club; Never-wet**
Orontium aquaticum L.
Araceae (Arum Family)
Flowers: March–April
Description: Emergent, perennial herb from a thick rhizome; 8" to 2' tall; flowers in a spadix; leaves in a basal cluster, either extending above or floating on water.
Range-Habitat: Scattered throughout the state but most abundant in the coastal plain; in the coastal plain common in muddy sites in alluvial and nonalluvial swamp forests, edge of ponds, tidal freshwater marshes, and ditches.
Comments: The common name "never-wet" alludes to the fact that leaves, when submerged, will come out of the water dry. Water drops will quickly roll of the leaf surface.
All parts of the plant contain crystals of calcium oxalate which may irritate the lining of the throat and cause it to swell—resulting in asphyxiation. Indians made use of the roots and seeds for food by boiling them extensively which dissolved the crystals.

139 **Red Maple**
Acer rubrum L.
Aceraceae (Maple Family)
Fruits: March–May
Description: Medium to large, deciduous tree; leaves opposite, usually three-to five-lobed; flowers perfect, or often male and female in separate clusters on the same or different trees; fruit a samara.
Range-Habitat: Common throughout the state in alluvial and nonalluvial swamp forests, beech and oak-hickory forests, pine–mixed hardwood forests, fence rows, and thickets.
Comments: The flowers bloom in our area sometimes as early as January, long before the leaves appear. The flowers are followed by the conspicuous winged fruits (samaras), bright red to yellow which produce a prominent display of color against the spring sky. Red maple is one of the pioneer trees in aquatic succession in abandoned rice fields and can establish itself in recently cleared, upland sites.
Red maple is often planted as a shade tree; the wood is used for a variety of products and maple syrup can be made from the sap, although the sap contains less sugar than the sugar maple.

140 **Butterweed**
Senecio glabellus Poiret
Asteraceae (Aster or Sunflower Family)
Flowers: March–June

Description: Annual, erect herb with smooth, hollow stems to 3' tall; often forming dense stands; leaves alternate, deeply divided into narrow segments.
Range-Habitat: Common throughout the piedmont and coastal plain; in the coastal plain it occurs in alluvial and nonalluvial swamp forests, hardwood bottoms, and wet pastures.
Comments: Worldwide there are twelve hundred species of the genus *Senecio,* and so far at least twenty-five have proven poisonous to livestock and humans. *S. glabellus* has been suspected of poisoning cattle in Florida. Refer to Kingsbury (1964) for an historical account of the genus *Senecio.*

141 **Leucothoe**
Leucothoe racemosa (L.) Gray
Ericaceae (Heath Family)
Flowers: March–Early June
Description: Deciduous shrub to 13' tall; racemes usually curved.
Range-Habitat: Chiefly found in the coastal plain and lower piedmont; in the coastal plain in all types of swamp forests, stream banks, pocosins, and savannas.

142 **Supplejack**
Berchemia scandens (Hill) K. Koch
Rhamnaceae (Buckthorn Family)
Flowers: April–May
Description: Woody, high-climbing, clockwise-twining vine with main stem to 7" thick; stems smooth, lustrous, green to reddish brown; leaves alternate, deciduous, with conspicuous veins; drupes blue-black when mature in August–October.
Range-Habitat: Common throughout the coastal plain; grows in alluvial and nonalluvial swamp forests, stream banks, and hardwood bottoms.
Comments: The drupes are an important food source for birds.

143 **Cross Vine**
Anisostichus capreolata (L.) Bureau
Synonym: *Bignonia capreolata* L.
Bignoniaceae (Trumpet-creeper Family)
Flowers: April–May
Description: Perennial, high-climbing, woody vine; leaves opposite, pinnately compound, with terminal leaflet modified into a tendril.
Range-Habitat: Very common throughout the state; in the coastal plain it occurs in alluvial swamp forests, all upland wooded habitats, fence rows, thickets, and disturbed areas.
Comments: The terminal leaflet that is modified into a tendril is highly branched. At the ends of the branches are small adhesive disks for attachment which allow cross vine to

climb the sides of buildings. The common name comes from the anatomy of the stem which is an easy aid to identification. A thin layer of barklike tissue separates the stem longitudinally into four equal segments which can be seen in a cross section.

Cross vine is a native vine that has spread widely into disturbed habitats.

144 **Blue Flag Iris**
Iris virginica L.
Iridaceae (Iris Family)
Flowers: April–May
Description: Perennial, erect herb, from short rhizome; stems 20" to 3' tall; leaves two-ranked, all flattened into one plane, at least basal.
Range-Habitat: Common throughout the coastal plain, rare in the piedmont; in the coastal plain it grows in alluvial and nonalluvial swamp forests, freshwater marshes, and ditches.
Comments: Various species of *Iris* are been found to be poisonous, containing an irritant principle in their leaves and rhizomes.

145 **Swamp Dogwood; Stiffcornel Dogwood**
Cornus stricta Lam.
Cornaceae (Dogwood Family)
Flowers: April–May
Description: Deciduous shrub or small tree to 15' tall; leaves opposite, lance-elliptic; twigs with white pith; drupe bright blue at maturity in July–August.
Range-Habitat: Throughout the state, but more common in the coastal plain; common in alluvial and nonalluvial swamp forests, hardwood bottoms, and stream and river banks.
Comments: Swamp dogwood has no commercial value because of its size; it is cultivated as an ornamental.

146 **Snowbell; Storax**
Styrax americana Lam.
Styracaceae (Storax Family)
Flowers: April–June
Description: Commonly a shrub, rarely a tree; deciduous, alternate, obovate leaves; flowers fragrant.
Range-Habitat: Chiefly a coastal plain species; common in alluvial and nonalluvial swamp forests, stream banks, and pocosins.

147 **Virginia Willow**
Itea virginica L.
Saxifragaceae (Saxifrage Family)
Flowers: May–June
Description: Deciduous shrub 3–6' tall; leaves alternate; inflorescence a terminal, narrow raceme.
Range-Habitat: Common throughout the state; in the coastal

plain found in alluvial and nonalluvial swamp forests, pond cypress swamps, and hardwood bottoms.

148 **Water Willow**
Justicia ovata (Walter) Lindau
Acanthaceae (Acanthus Family)
Flowers: May–July
Description: Colonial, rhizomatous, perennial herb, 4–20" tall; leaves opposite; flowers in spikes from long, axillary stalks.
Range-Habitat: A coastal plain species; common in alluvial and nonalluvial swamp forests, hardwood bottoms, and pond and stream margins.

149 **Lizard's Tail**
Saururus cernuus L.
Saururaceae (Lizard's Tail Family)
Flowers: May–July
Description: Perennial herb, often forming extensive colonies by rhizomes; aquatic or terrestrial; spike drooping in flower, but becoming erect in fruit.
Range-Habitat: Common throughout the coastal plain and piedmont; in the coastal plain it grows in a variety of aquatic habitats including alluvial and nonalluvial swamp forests, streams, lakes and pond margins, low woodlands, and ditches.
Comments: The common name comes from the drooping spike.

150 **Bur-reed**
Sparganium americanum Nuttall
Sparganiaceae (Bur-weed Family)
Flowers: May–September
Description: Perennial herb; erect, 1–3' tall; stems zigzag; leaves alternate, two-ranked, sheathing at the base; male and female flowers in separate heads; smaller male heads above, female heads below; male flowers wither and die as soon as pollen is shed.
Range-Habitat: Chiefly coastal plain and mountains; in the coastal plain found in open muddy areas of swamp forests, streams, roadside ditches, and shallow ponds.
Comments: Bur-weed is an emergent plant that sometimes forms dense stands. The seeds are eaten by waterfowl and marsh birds.

151 **Ladies'-eardrops**
Brunnichia cirrhosa Banks ex Gaertner
Polygonaceae (Buckwheat Family)
Flowers: June–July
Description: Partly woody vine climbing by tendrils; leaves alternate, ovate; calyx of fruit modified into winglike structure; fruits mature August–September.
Range-Habitat: An outer coastal plain species; rare in alluvial swamp forests and along river banks.
Comments: Ladies'-eardrops is a little-known vine to botanists

and laypeople. I, however, have observed it many times, most often in openings on the edge of swamps and along river banks where it forms dense mats on low vegetation. The common name comes from the shape of the winged fruit.

152 **Aquatic Milkweed**
Asclepias perennis Walter
Asclepiadaceae (Milkweed Family)
Flowers: June–August
Description: Perennial herb with milky juice; leaves opposite; corolla brilliant white; mature fruits hang down, mature August–September.
Range-Habitat: A coastal plain species; common in alluvial and nonalluvial swamp forests and hardwood bottoms.

153 **Green-fly Orchid**
Epidendrum conopseum R. Brown
Orchidaceae (Orchid Family)
Flowers: July–September
Description: Epiphyte; stems erect or ascending, smooth, to 16" tall; leaves evergreen, leathery; roots with well-developed, whitish velamen; flowers fragrant.
Range-Habitat: Evidently an outer coastal plain species; common in alluvial swamp forests and hardwood bottoms; also, occasional on live oak trees in various sites throughout the coastal plain.
Comments: This is the only epiphytic orchid in South Carolina. Based on my field observations, it is common rather than rare as reported by most manuals. It is often overlooked because it grows high up trees where it is sometimes hidden in growth of resurrection fern and/or Spanish moss. Populations have been found in almost every river system in the coastal plain. It generally grows on live oak, bald cypress, and tupelo gum.

154 **Macbridea; Carolina Birds-in-a-Nest**
Macbridea caroliniana (Walter) Blake
Lamiaceae (Mint Family)
Flowers: July–August
Description: Perennial herb to 3' tall; flowers in one to three tight, bracted, separated clusters; leaves opposite, in seven to eleven pairs.
Range-Habitat: A rare coastal plain species; in openings in swamp forests, freshwater marshes and bogs.
Comments: The common name, birds-in-a-nest, comes from the flowers (birds) that arise out of the empty bracts (the nest).

155 **Elderberry**
Sambucus canadensis L.
Caprifoliaceae (Honeysuckle Family)
Fruits: July–August

Description: Deciduous shrub or occasionally small tree reaching 20' tall; often forming thickets due to production of new shoots from the root system; year-old twigs with prominent lenticles; pith of second year or older wood white; leaves opposite, odd-pinnately compound with five to eleven leaflets; berries purplish to blackish, mature July–August.

Range-Habitat: Common throughout the state; in the coastal plain in hardwood bottoms, alluvial and nonalluvial swamp forests, roadside ditches, and pastures—usually in open places.

Comments: Elderberry is a versatile plant. The fruits make excellent jelly, jam, pies, and wine; they are an important food for game and songbirds. The soft pith is easily removed so the stem is sometimes used to make flutes and whistles. The plant is poisonous except for the cooked, ripe berries. The unripe berries and uncooked ripe berries cause nausea. Children have been poisoned by using the stems as peashooters.

156 **False Nettle**

Boehmeria cylindrica (L.) Swartz

Urticaceae (Nettle Family)

Flowers: July–August

Description: Perennial herb without stinging hairs; stems 1.5–3' tall; leaves opposite, long-stalked; tiny flowers are in small, headlike clusters, arranged in continuous (female) or interrupted (male) spikes in the axils of opposite leaves; spikes often terminated by leaves.

Range-Habitat: Common throughout the state; in the coastal plain on fallen logs, stumps, cypress knees, and buttresses in alluvial and nonalluvial swamp forests; also found in hardwood bottoms, freshwater marshes, and low ground.

157 **Skullcap**

Scutellaria lateriflora L.

Lamiaceae (Mint Family)

Flowers: July–Frost

Description: Perennial, smooth herb with four-angled stems; erect stems 1–5' tall, freely branched; creeping stems present; leaves opposite.

Range-Habitat: Common throughout the state; in the coastal plain in alluvial and nonalluvial swamp forests, on floating logs, cypress knees, buttresses, and stumps; also in hardwood bottoms and freshwater marshes.

158 **Purple Lobelia**

Lobelia elongata Small

Campanulaceae (Bellflower Family)

Flowers: August–October

Description: Perennial herb from basal offshoots; to 5' tall; leaves alternate, narrowly lanceolate, tapering to both ends.

Range-Habitat: Common throughout the coastal plain; grows in alluvial and nonalluvial swamp forests, freshwater marshes, and longleaf pine savannas.

Comments: Kingsbury (1964) gives a detailed account of the common lobelias; he does not mention purple lobelia. He does say that species of *Lobelia* indigenous to America were used in a variety of medicines, but that cases of death from overdoses of medicinal preparations were frequent. Purple lobelia, then, should be considered poisonous until proven otherwise.

159 **Poison Ivy**

Rhus radicans L.

Synonym: *Toxicodendron radicans* (L.) Kuntze

Anacardiaceae (Cashew Family)

Fruits: August–October

Description: High-climbing vine with numerous, adventitious roots; trailing or upright; leaves trifoliolate, the lateral leaflets oblique, entire, or remotely toothed; flowers mature late April–May.

Range-Habitat: Common throughout the state; in the coastal plain it is found in swamp forests and most wooded habitats, fence rows, meadows, and disturbed sites.

Similar Species: A similar plant is poison oak (*Rhus toxicodendron* L.). The distinction between poison oak and poison ivy is uncertain. Poison oak may be only varietally different or it may be an ecotype. Poison oak is an erect plant growing in sandy woodlands. It is as strong a contact irritant as poison ivy.

Comments: All parts of poison ivy contain an oleoresin (urushiol) that can cause severe skin inflammation, itching, and blistering on contact. The oleoresin can also be contacted when it is borne on smoke or dust particles, clothes, and the hair of animals.

Poison ivy is a highly variable species. It can occur as a ground cover in yards or roadsides, or as a large vine. Its fruits are eaten by many songbirds with no ill effects.

160 **Tupelo Gum; Water Tupelo**

Nyssa aquatica L.

Nyssaceae (Tupelo Tree Family)

Fruits: September–October

Description: Large deciduous tree, often over 100' tall; pronounced buttress develops when growing in water; leaves alternate, entire or irregularly and sparsely toothed; male and female flowers on same plant; male flowers clustered in heads, female flowers solitary; fruit a drupe.

Range-Habitat: A common coastal plain species; occurs in alluvial and nonalluvial swamp forests.

Comments: The drupes are eaten by a wide variety of wildlife. Francis P. Porcher (1869) reported that the roots are white, spongy, and light, and were sometimes used in the southern states as a substitute for cork. He also said that he had shoes made of tupelo wood for slaves on a plantation. Perhaps the most noted use of tupelo is for honey which is sold in specialty stores.

161 **Coral Greenbrier**
Smilax walterii Pursh
Liliaceae (Lily Family)
Fruits: September–Winter
Description: High-climbing, deciduous, woody vine; spines lacking; berries bright red, persisting through the winter.
Range-Habitat: Common throughout the coastal plain, rare in the outer piedmont; in the coastal plain it grows in alluvial and nonalluvial swamp forests and hardwood bottoms.
Comments: This is the only deciduous greenbrier in the Lowcountry. In leaf it is difficult to separate from other greenbriers. The leafless vines with berries are often used for Christmas decorations.

162 *Hardwood Bottoms*

163 **Jack-in-the-pulpit; Indian Turnip**
Arisaema triphyllum (L.) Schott
Araceae (Arum Family)
Flowers: March–April
Description: Erect, perennial herb from a corm, 8–30" tall; leaves one or two, palmately divided; flowers on a fleshy spadix with male above, female below; spathe (the pulpit) with a tube and a hood which arches over the spadix (Jack); fruit a red berry in clusters, maturing in July.
Range-Habitat: Common throughout the state; in the coastal plain it occurs in hardwood bottoms and low areas in oak-hickory and beech woods.
Comments: All parts of the plant, but especially the corm, contain crystals of calcium oxalate that can irritate the mucous membranes of the mouth and throat causing a burning sensation. Death can result by asphyxiation if the air passages swell. Fernald and Kinsey (1958) report, however, that the crystals can be broken up by heat and drying (but not by boiling). Once done, the corm becomes mild and pleasant tasting. It was used by Native Americans as flour—hence, the common name Indian turnip.

164 **Pawpaw**
Asimina triloba (L.) Dunal
Annonaceae (Pawpaw Family)
Flowers: March–May
Description: Large shrub or small tree, 16–33' tall; twigs covered with fine rust-colored hairs; leaves deciduous, alternate, simple, malodorous when crushed; flowers borne on the

wood produced the previous year; winter buds flattened, covered with rust-colored hairs; fruits mature in fall.
Range-Habitat: Throughout the state but probably only occasional in the coastal plain; occurs in hardwood bottoms in the coastal plain, often forming dense thickets.
Comments: The ripe fruit of pawpaw is sweet; it can be eaten raw, baked as pie filling, or made into a variety of other foods. The fruits are collected when green (often from the ground) and kept until ripe. The fruits are readily eaten by wildlife. Early settlers made yellow dye from the ripe pulp.

165 **Easter Lily; Naked Lady; Atamasco Lily**
Zephyranthes atamasco (L.) Herbert
Amaryllidaceae (Amaryllis Family)
Flowers: Late March–April
Description: Perennial herb from a bulb; flowering stalk to 1' tall, generally solitary, terminated by a single flower; leaves basal, linear; perianth usually white, rarely pink.
Range-Habitat: Common in the coastal plain and lower piedmont; in the coastal plain in hardwood bottoms and wet meadows.
Comments: Leaves, and especially the bulb, are highly poisonous to horses, cattle, and fowl (and may also be poisonous to people). The common name, naked lady, comes from the leafless stalk terminated by the beautiful flower.

166 **Dog-hobble**
Leucothoe axillaris (Lam.) D. Don
Ericaceae (Heath Family)
Flowers: Late March–May
Description: Evergreen shrub to 5' tall, with clustered, arching stems usually forming dense colonies.
Range-Habitat: Chiefly a coastal plain species; common on edges of alluvial and nonalluvial swamp forests, in pocosins, and along stream banks.
Comments: This common name comes from dense colonies which are almost impenetrable—even for a hunting dog.

167 **Arrowwood**
Viburnum dentatum L.
Caprifoliaceae (Honeysuckle Family)
Flowers: Late March–May
Description: Deciduous shrub, 3–15' tall; twigs hairy; flowering stems with one to two pairs of opposite, coarsely toothed leaves; drupe blue to black, maturing July–September.
Range-Habitat: In South Carolina primarily a coastal plain species; common in hardwood bottoms and alluvial and nonalluvial swamp forests.
Comments: This is one of the many similar viburnams with edible fruit. The pulp of the fruit is sweet.

168 **Leather-flower**
Clematis crispa L.
Ranunculaceae (Buttercup Family)
Flowers: April–August
Description: Perennial, herbaceous vine, climbing or weakly ascending; leaves opposite, pinnately compound with three to five leaflets; flowers extremely fragrant.
Range-Habitat: Chiefly coastal plain; common in hardwood bottoms, freshwater marshes, and alluvial and nonalluvial swamp forests.

169 **Fever Tree; Georgia Bark**
Pinckneya pubens Michaux
Rubiaceae (Madder Family)
Flowers: May
Description: Small tree or shrub; leaves deciduous, opposite, and simple; flowers borne in clusters; petals greenish yellow with bright red spots on the inside; flower made conspicuous by enlargement of one (and often two) calyx lobe which is bright rose pink with greenish veins.
Range-Habitat: Very rare tree in South Carolina and known in only Beaufort and Jasper counties (occurs only in South Carolina, Georgia, and Florida); outer coastal plain along the edges of hardwood bottoms or swamp forests.
Comments: Fever tree has long been reported from sites in Beaufort County, especially near Bluffton, by early botanists such as Mellichamp. Recently I found it in Jasper County. It is certainly one of the more interesting plants in South Carolina from a historical and biological position. It is only conspicuous for a short time in May when in bloom.

The common name, fever tree, comes from the use of a chemical in the bark as a substitute for quinine in treating malaria.

170 **Climbing Hydrangea**
Decumaria barbara L.
Saxifragaceae (Saxifrage Family)
Flowers: May–June
Description: High-climbing, evergreen, woody vine with opposite leaves; aerial roots are adhesive at their tips.
Range-Habitat: Common throughout the state; in the coastal plain found in alluvial and nonalluvial swamp forests and hardwood bottoms.
Comments: Climbing hydrangea is a spectacular vine in flower, but it is not commonly noticed because the flowers are high off the ground. It is easily identified since it is our only woody vine that has opposite, entire leaves and climbs by aerial roots.

171 **Cow-itch; Trumpet Vine**
Campsis radicans (L.) Seemann
Bignoniaceae (Trumpet-creeper Family)
Flowers: June–July
Description: Deciduous, woody vine, trailing or high-climbing by

means of two, short rows of aerial roots from the nodes; leaves opposite, pinnately compound with seven to fifteen leaflets.
Range-Habitat: Common throughout the state; in the coastal plain in hardwood bottoms, all other forested communities, fence rows, vacant lots, roadsides, and yards.
Comments: Cow-itch is a native vine that has exploited disturbed habitats throughout its range. It is often cultivated for its attractive flowers. Contact with cow-itch may cause skin inflammation and blisters in sensitive people.

172 **River Oats**
Uniola latifolia Michaux
Poaceae (Grass Family)
Flowers: June–October
Description: Rhizomatous, perennial herb; stems 2–5' tall.
Range-Habitat: Throughout the state; in the coastal plain in hardwood bottoms, along river and stream banks, and ditches.
Comments: River oats is sometimes planted as an ornamental; it is also dried and used for bouquets and floral arrangements.

173 **Sweet-gum; Red-gum**
Liquidambar styraciflua L.
Hamamelidaceae (Witch-hazel Family)
Fruits: Fall–Winter
Description: Large, deciduous tree; branches frequently irregularly corky-winged; leaves alternate, palmately five-lobed; male and female flowers in separate clusters on the same plant; seed capsules in hard, rounded ball-like structures which persist throughout the winter.
Range-Habitat: Common throughout the state; in the coastal plain in hardwood bottoms, pine–mixed hardwood forests, longleaf pine flatwoods, beech woods, abandoned fields, and thickets.
Comments: Sweet-gum is one of the most versatile trees. Although it is not strong enough for structural timber, its pink or ruddy heartwood shows handsome figures on the quarter-sawed cut and is used for veneer, furniture, and plywood panels. Today, sweet-gum accounts for a higher commercial harvest than any other deciduous hardwood.

During pioneer times in the South, the gum was used for treatment of sores and skin troubles, for a chewing gum, and in treatment of dysentery by the Confederate armies. During World War I and II, its gum was the base of soaps, drugs, and adhesives.

Its seeds are of minor importance to wildlife.

THE DECIDUOUS FOREST GARDENS

174 *Marl Forests*

175 **Blackstem Spleenwort**
Asplenium resiliens Kunze
Aspleniaceae (Spleenwort Family)

Description: Evergreen fern; leaves with black rachis and leaf stalks; leaflets mostly opposite, entire, auricled on the upper side.
Range-Habitat: A rare fern from scattered localities in the coastal plain and mountains; it grows in rather dry crevices on shaded limestone and marl outcrops.
Comments: This fern is rare because of the limited sites of shaded limestone and marl outcrops.
Similar Species: A similar species is *Asplenium heteroresiliens* Wagner which grows in the same habitat. It can be distinguished by its shallowly toothed leaflets. Wagner's spleenwort is very rare and known from only a few sites in the coastal area.

176 **Meadow Parsnip**
Thaspium barbinode (Michaux) Nuttall
Apiaceae (Parsley Family)
Flowers: April–May
Description: Erect, herbaceous, perennial herb from a thickened, fibrous root; 2–4' tall; all leaves two to three times divided (with divisions in threes).
Range-Habitat: Chiefly piedmont and mountains; infrequent in the coastal plain in marl forests and rich beech and oak-hickory woods.

177 **Alumroot**
Heuchera americana L.
Saxifragaceae (Saxifrage Family)
Flowers: April–June
Description: Rhizomatous perennial with long leaf stalks and rounded blades; 2–3' tall; stamens with orange anthers that protrude from the flowers.
Range-Habitat: Primarily a mountain and piedmont species; rare in the coastal plain in several widely scattered marl forests in Dorchester and Berkeley counties.

178 **Thimbleweed**
Anemone virginiana L.
Ranunculaceae (Buttercup Family)
Flowers: May–July
Description: Herbaceous, hairy perennial, from a short, thick rootstock or rhizome; often growing in colonies; 4–8" tall; petals absent and the sepals showy white.
Range-Habitat: Primarily a mountain and piedmont species; occurs rarely in the coastal plain in marl forests and reported from Berkeley and Dorchester counties.
Comments: The thimble-shaped cluster of pistils is the distinctive feature of this species and accounts for the common name.

179 **Elytraria**
Elytraria caroliniensis (J. F. Gmelin) Persoon
Synonym: *Tubiflora caroliniensis* (Walter) J. F. Gmelin

Acanthaceae (Acanthus Family)
Flowers: June–August
Description: Perennial herb from elongate, slender rhizomes; flowering stalks 4–20" tall; leaves in a basal cluster.
Range-Habitat: Infrequent coastal plain species; reported in Berkeley, Colleton, Dorchester, and Hampton counties; swamp forests and marl forests.
Comments: Elytraria is listed as rare by most manuals; however, I believe it is more common than reported as I have collected and observed it from numerous sites throughout the coastal area.

180 **Crested Coral-root**
Hexalectris spicata (Walter) Barnhart
Orchidaceae (Orchid Family)
Flowers: July–August
Description: Saprophytic, perennial herb, 6–30" tall; stem flesh colored, lacking in chlorophyll; leaves reduced to sheathing, scalelike bracts.
Range-Habitat: Widely scattered locations throughout the state; infrequent in the coastal plain in calcareous or circumneutral soils such as Indian middens and marl forests; also in rich beech and oak-hickory forests.
Comments: This is the most attractive of the saprophytic orchids as the stem supports a spike of richly colored flowers. Like most saprophytes, it is almost impossible to successfully transplant.

181 **White Basswood**
Tilia heterophylla Vent
Tiliaceae (Basswood Family)
Fruits: July–August
Description: Deciduous, medium-sized tree; flowers in clusters of ten to twenty, attached by a slender stalk to a narrow, leaflike bract, flowering in June; nutlike fruits remain attached to the bract until shed in late winter or early spring.
Range-Habitat: Chiefly mountains and piedmont; rare and scattered widely in the coastal plain in marl forests, beech forests, and Indian middens.

182 **Horse-balm**
Collinsonia tuberosa Michaux
Lamiaceae (Mint Family)
Flowers: Late July–September
Description: Perennial herb with tuberous roots; erect, 12–30" tall; stems four-angled; leaves opposite, usually four to twelve pairs at flowering and fruiting.
Range-Habitat: Infrequent and scattered throughout the piedmont and coastal plain; rich woods and marl forests.

183 **Hop Hornbeam**
Ostrya virginiana (Miller) K. Koch

Betulaceae (Birch Family)
Fruits: August–October
Description: Small, deciduous tree with brown, shredding bark; occasionally reaching 35–50' tall; male and female catkins produced in April–May; mature female catkin consists of nutlets, each enclosed by membranous inflated sacs arranged in a spike as shown in the photograph.
Range-Habitat: Chiefly piedmont and mountains; rare in the coastal plain where it appears to be confined to marl forests.
Comments: Hornbeam is probably more common in the coastal plain than reported as I have collected and observed it from numerous sites. Hop hornbeam is usually an understory species and not large enough to be of commercial importance.

184 **Buckthorn**
Rhamnus caroliniana Walter
Rhamnaceae (Buckthorn Family)
Fruits: September–October
Description: Slow-growing, short-lived shrub or small tree, 30–40' tall; leaves deciduous, but often remaining through the winter; parallel veins from the midrib of the leaf are prominent.
Range-Habitat: Chiefly in the piedmont; scattered in the coastal plain in marl forests and shell mounds along the coast.
Comments: The sweet fruits are eaten by many birds. The wood is of no commercial value.

185 ***Beech Forests***

186 **Bloodroot**
Sanguinaria canadensis L.
Papaveraceae (Poppy Family)
Flowers: Early March–April
Description: Perennial herb, 4–16" high, with no leafy stem; rhizomes with bright orange-red juice; flower single on a leafless stem, opening in the day, closing at night, and lasting only a short time; the single leaf continues to enlarge after the petals drop.
Range-Habitat: Chiefly mountains and piedmont; scattered in the outer coastal plain but more common than reported; in beech forests and rich oak-hickory forests.
Comments: Indians used the red juice from the roots to dye baskets and clothing and to make war paint. Appalachian crafters today still use the red juice to dye baskets and cloth. Indians also used the sap as an insect repellent and used the dried root to cure rattlesnake bites.

When it became known that the Indians used the root as a somewhat successful treatment for tumors, interest in the plant increased. Drugs developed from bloodroot effectively treat ringworm and eczema.

187 **Blue Star**
Amsonia tabernaemontana Walter
Apocynaceae (Dogbane Family)
Flowers: March–April
Description: Perennial herb 12–28" tall; usually several stems from a woody rootstock; leaves simple, entire, alternate; juice milky.
Range-Habitat: Chiefly piedmont; rare and scattered in the coastal plain in beech forests.

188 **Pennywort**
Obolaria virginica L.
Gentianaceae (Gentian Family)
Flowers: March–April
Description: Fleshy perennial, 3–6" high; roots brittle; leaves opposite; flowers usually in groups of three in the axils of purplish, bractlike leaves.
Range-Habitat: Throughout the mountains and piedmont; rare in the coastal plain in beech forests; often overlooked because of its small size, especially during early growth.

189 **May-apple; Mandrake**
Podophyllum peltatum L.
Berberidaceae (Barberry Family)
Flowers: March–April
Description: Smooth, rhizomatous, perennial herb 12–18" tall; solitary, nodding flower from the base of a pair of large, deeply lobed leaves; leaves deciduous, with the leaf stalk attached in the center on the underside of the leaf; berry yellow or red when ripe in May–June.
Range-Habitat: Common throughout the state; in the coastal plain in beech forests, rich oak-hickory forests, meadows, and moist road banks.
Comments: Today, alkaloids from the rhizomes are being used as the basis of a drug to treat lymphocytic leukemia and testicular cancer. The plant has long been known to contain antitumor alkaloids. Indians used extracts of the rhizomes as purgatives and for skin disorders and tumorous growths.

The leaves, rhizomes, and seeds are poisonous if eaten in large quantities; ripe fruits can be made into marmalade.

Often single-leaved plants that never develop flowers are observed.

190 **Little Sweet Betsy**
Trillium cuneatum Raf.
Liliaceae (Lily Family)
Flowers: Mid-March–April
Description: Perennial herb from rhizome; stems 4–12" tall;

leaves simple, three in a whorl, mottled with lighter green and purple; single, stalkless, ill-scented flower; petals and sepals purple, maroon, or wine red.
Range-Habitat: Frequent in the mountains, southern and central piedmont, and southern coastal plain; in the coastal plain it grows in beech and marl forests.
Comments: Sweet Betsy likes circumneutral to basic soil. I have collected little sweet betsy on Indian middens at the mouth of the North Edisto River in Charleston County within a few feet of the salt marsh.

191 **Heart-leaf**
Hexastylis arifolia (Michaux) Small
Aristolochiaceae (Birthwort Family)
Flowers: March–May
Description: Rhizomatous, perennial herb with no above-ground stem; leaves aromatic, evergreen, often in clusters; flowers without petals; calyx flask-shaped; flowers often hidden under leaf litter.
Range-Habitat: Common throughout the state; in the coastal plain occurs in beech, oak-hickory, and marl forests and swamp forests.

192 **Green-and-gold**
Chrysogonum virginianum L.
Asteraceae (Aster or Sunflower Family)
Flowers: Late March–Early June
Description: Fibrous-rooted, herbaceous perennial, flowering early when very small; at first stemless or short-stemmed, but later flowering stems elongating to 1–2' tall; leaves opposite.
Range-Habitat: Common throughout the state; in the coastal plain it occurs in beech and oak-hickory forests.

193 **Cancer-root; Squaw-root**
Conopholis americana (L.) Wallroth
Orobanchaceae (Broomrape Family)
Flowers: March–June
Description: Herbaceous holoparasite, on roots of the red oak group; 3–10" tall; leaves reduced to brown scales; entire plant pale brown or yellow brown.
Range-Habitat: Common throughout the state; in the coastal plain in beech and oak-hickory forests.
Comments: This plant starts its life cycle as a underground gall-like mass (combination of host and parasite tissue) on small oak roots. It takes about five years for the underground structure to mature, after which it produces above-ground flowering stems for many years. The stems dies back in June, but the dried remains can be seen for months.

194 **Southern Twayblade**
Listera australis Lindley
Orchidaceae (Orchid Family)

Flowers: March–July
Description: Inconspicuous herb, 3–14" tall; leaves in twos, opposite, dark green, at the middle of the stem.
Range-Habitat: A coastal plain species; common but often overlooked in beech and rich oak-hickory forests.
Comments: Some manuals list this plant from pine woods, but I have never observed it in pine woods in the coastal area. Most manuals list it as rare; however, it is my experience that it is common, but overlooked.

195 **Red Buckeye**
Aesculus pavia L.
Hippocastanaceae (Buckeye Family)
Flowers: April–Early May
Description: Deciduous shrub or small tree, seldom over 20–25' tall; leaves opposite and palmately lobed with five to seven leaflets; capsule matures July–August and contains two large seeds that turn dark brown at maturity.
Range-Habitat: Chiefly a coastal plain species; common in beech and rich oak-hickory forests, marl forests, Indian middens, and edges of swamp forests.
Comments: All parts of the plant contain a glycoside, called aesculin, that is highly poisonous. The practice of making a necklace of the shiny seeds is dangerous since the wearer may suck on the seeds. Indians would make doughballs from the seeds and toss them in a river where fish would eat them. The fish would be stunned by the poison and float to the surface where they were harvested.

196 **Violet Wood Sorrel**
Oxalis violacea L.
Oxalidaceae (Wood Sorrel Family)
Flowers: April–May; Fall
Description: Perennial herb, 4–8" tall; leaves and flowering stalks all arise from a bulbous base; leaves palmately divided into three segments, reddish or purple beneath.
Range-Habitat: Common throughout the state; in the coastal plain in beech and oak-hickory forests.
Comments: Some sources report that wood sorrel flowers in the fall after the leaves die. Whether it flowers in the fall in the coastal plain is not documented. In small amounts the leaves make a refreshing addition to a salad; large quantities should not be eaten because of the presence of oxalic acid. It is often cultivated.

197 **Wild Ginger**
Asarum canadense L.
Aristolochiaceae (Birthwort Family)
Flowers: April–May
Description: Rhizomatous, perennial herb, growing at ground

level; leaves opposite, in pairs on a stem; deciduous; flower solitary, arising between the two leaf stalks; flowers often inconspicuous, hidden by the forest litter.

Range-Habitat: Scattered in all three provinces in the state; based on my field observations, it is more widespread in the coastal plain than recorded; infrequent in beech and marl forests.

Comments: Wild ginger is a plant of many uses. The rhizomes have a strong odor similar to true ginger and are often used as a substitute. Indians made candy of the rhizomes by boiling them until tender, then dipping in syrup. Numerous sources suggest, however, that consumption of excessive amounts of *Asarum* should be avoided because it may have cancer-causing properties.

198 **Sweet Shrub**

Calycanthus floridus L.

Calycanthaceae (Strawberry-shrub Family)

Flowers: April–May

Description: Deciduous, aromatic shrub 3–10" tall; leaves opposite.

Range-Habitat: Chiefly a mountains and piedmont species; probably infrequent in the inner coastal plain and rare in the outer coastal plain; in the outer coastal plain it occurs in beech and oak-hickory forests and along stream banks.

Comments: Sweet shrub is widely planted as an ornamental because of the spicy fragrance of its flowers; its seeds are poisonous.

199 **Bellwort; Perfoliate Bellwort**

Uvularia perfoliata L.

Liliaceae (Lily Family)

Flowers: April–May

Description: Perennial, colonial, erect herb, 6–16" tall; stems unbranched or one-branched, arising from slender, white, underground stolons; leaves alternate, perfoliate.

Range-Habitat: Chiefly a common mountains and piedmont species; rare or local in the coastal plain in beech and rich oak-hickory forests.

Comments: The genus name, *Uvularia,* comes from uvula—the soft lobe hanging into the throat from the soft palate. It was once thought because of the drooping flowers that it could be used to treat throat diseases.

200 **Tulip Tree; Yellow Poplar**

Liriodendron tulipifera L.

Magnoliaceae (Magnolia Family)

Flowers: April–June

Description: Large, straight-trunked tree, 98–165' tall; deciduous; winter buds flattened and enclosed by two, nonoverlapping scales.

Range-Habitat: Common throughout the state; in the coastal plain it occurs in beech and rich oak-hickory forests, nonalluvial swamp forests, and disturbed places.

Comments: Tulip trees of the original forest grew to 165' tall. Trees grown in forest conditions were 80–100' tall before producing branches, a feature that made it a superior lumber tree. Tulip tree wood was used for lumber more than any other tree except for pine. Today its wood has a myriad of uses: interior finishes, furniture, general construction, and plywood. The Charleston furniture maker Thomas Elfe (1719–1775) employed it in his works. The tulip tree is also planted as a shade tree. The seeds are an important food for wildlife.

201 **Indian Cucumber-root**

Medeola virginiana L.

Liliaceae (Lily Family)

Flowers: April–June

Description: Perennial herb from a white rhizome; stems 1–2.5' tall; leaves usually in two whorls with the upper whorl generally of three leaves and the lower of six to ten leaves.

Range-Habitat: Scattered throughout the state; infrequent in the coastal plain in beech and oak-hickory forests.

Comments: Native Americans used the rhizomes for food; they are crisp and starchy, with a taste like cucumber.

202 **Wild Geranium; Spotted Cranesbill**

Geranium maculatum L.

Geraniaceae (Geranium Family)

Flowers: April–June

Description: Herbaceous perennial from thick rhizome; stems 8–24" tall; basal leaves with long stalks; fruit with a long beak—hence, the common name cranesbill.

Range-Habitat: Chiefly mountains and piedmont; scattered in the coastal plain in beech forests.

Comments: Native Americans made a powder from the dried rhizome and used it as a styptic and astringent to slow bleeding from cuts.

The method of seed dispersal is unusual. At the base of the fruit are five small cups facing outward and attached to springy bands connected to the top of the fruit. As the bands dry, they come under tension; they break loose, and as they curl upward, the cups are flung outward, throwing the seeds several feet from the plant. One can touch the ripe, unopened fruits causing them to open.

203 **Spotted Wintergreen; Pipsissewa**

Chimaphila maculata (L.) Pursh

Ericaceae (Heath Family)

Flowers: May–June

Description: Small rhizomatous herb to 8" tall; leaves evergreen, whorled, and variegated with white veins on the midrib and larger veins elsewhere; flowers nodding, fragrant.

Range-Habitat: Common throughout the state; in the coastal plain common in beech, oak-hickory, and pine–mixed hardwood forests.

Comments: Pipsissewa has a variety of medicinal uses. Native Americans used the plant to treat kidney stones and used a poultice from the leaves to heal blisters or reduce swelling in legs or feet. Early settlers used a tea as a febrifuge to break fevers from typhus. A popular refreshing hot tea was made by pouring boiling water over chopped leaves. Until recently it was used to flavor commercial root beer.

204 **Partridge Berry; Twin-flower**

Mitchella repens L.

Rubiaceae (Bedstraw Family)

Flowers: May–June

Fruits: June–Winter

Description: Trailing, creeping, evergreen perennial; leaves opposite; flowers in pairs; a single berry forms from the fusion of the ripened ovaries of two flowers, leaving two scars on the mature fruit; fruits mature in June–July, but persist throughout the winter.

Range-Habitat: Common throughout the state; in the coastal plain in a variety of wooded habitats including beech, maritime, and oak-hickory forests.

Comments: The berries are edible although bland and seedy. It makes an excellent groundcover for home gardens, especially under acid-loving shrubs. Indian women made a tea from the leaves as an aid in childbirth; it is used today in commercial preparations to help regulate the menstrual cycle in women no longer taking birth control pills.

205 **Silky Camellia; Virginia Stewartia**

Stewartia malacodendron L.

Theaceae (Tea Family)

Flowers: May–June

Description: Deciduous shrub or small tree, seldom over 20' high; leaves thin, alternate, deciduous, with fine, sharp-pointed teeth along the margin.

Range-Habitat: Uncommon in the coastal plain; rare in the lower piedmont; occurs as an understory species in scattered localities in rich, deciduous woods; most often along streams and creeks.

Comments: The flower of silky camellia resembles the cultivated camellias and is often cultivated as an ornamental. The trees are too small for lumber and of little or no value for wildlife.

206 **Lopseed**

Phryma leptostachya L.

Phrymaceae (Lopseed Family)

Fruits: August–October
Flowers: June–August
Description: Perennial herb with erect stems 1–3' tall; leaves opposite; raceme with flower buds erect; flowers horizontal to ascending and fruits hanging down.
Range-Habitat: Chiefly mountains and piedmont; infrequent in the coastal plain in beech and rich oak-hickory forests.
Comments: The downward-hanging fruits (hence, the common name) make this plant easy to identify. *Phryma* is the only genus in the family.

207 **Green Adder's Mouth**
Malaxis unifolia Michaux
Orchidaceae (Orchid Family)
Flowers: June–August
Description: Bright green, smooth herb with solitary, clasping leaf; plant 3–20" tall; stem from a bulbous corm.
Range-Habitat: A coastal plain species; common in beech forests, shaded edge of swamps and rich oak-hickory and pine–mixed hardwood forests.

208 **American Beech**
Fagus grandifolia Ehrhart
Fagaceae (Beech Family)
Fruits: September–October
Description: Large tree with smooth, gray bark; leaves deciduous, but often remaining throughout the winter; terminal buds long and slender, about 1" long; flowers in March–April; male and female flowers in separate clusters on the same tree; large fruit production occurs every two to three years.
Range-Habitat: Throughout the state; in the outer coastal plain it is common in beech forests where it is often the dominate tree, and occasional in rich oak-hickory forests.
Comments: Beech is a beautiful, long-lived tree relatively free from disease and ideal for cultivation. The oily seeds are an important wildlife food. Beech-nut oil has never been important in America, but in Europe beech-oil (from *Fagus sylvatica*) is an important commercial product. Whether American beech could yield sufficient oil has never been tested. The nut is one of the sweetest in North America; at one time it was sold by New England country groceries.

The wood has never been an important commercial product, although it did gain fame by being made into clothespins because of its elastic nature.

209 **Beech-drops**
Epifagus virginiana (L.) Barton
Orobanchaceae (Broomrape Family)
Flowers: September–November
Description: Holoparasite on the roots of beech trees; annual;

plant 4–18" tall, branched; dried stalks persisting throughout the winter; upper flowers sterile, lower flowers not opening and self-fertilized, producing abundant seeds; leaves reduced to scales.

Range-Habitat: Common throughout the state wherever beech trees occur.

Comments: The dustlike seeds are carried down through the soil by percolating water where they germinate in the rhizosphere of beech roots.

210 **Witch-hazel**

Hamamelis virginiana L.

Hamamelidaceae (Witch-hazel Family)

Flowers: October–December

Description: Large shrub or small tree which may reach heights of 30–35'; leaves alternate, deciduous, and scalloped along the margin; capsules mature October–November the following year.

Range-Habitat: Common throughout the state; in the coastal plain it occurs in beech and oak-hickory forests.

Comments: Witch-hazel is too small for lumber. Witch-hazel extract used as an astringent is made from the dormant stems and twigs. Experienced water diviners use branched twigs as a divining rod.

The four, narrow petals of witch-hazel curl back into a bud when the temperature falls, then expand again when it gets warmer. The seeds are ejected from the mature capsule with considerable force.

211 *Oak-Hickory Forests*

212 **Flowering Dogwood**

Cornus florida L.

Cornaceae (Dogwood Family)

Flowers: March–April

Description: Fast growing, short-lived tree to 50' tall; bark rough, fissured into small, thin, angular scales; leaves opposite, deciduous, turning scarlet in the fall; flowers in heads, surrounded by four large, white, petal-like bracts notched at the apex; berries scarlet-red, maturing in the fall.

Range-Habitat: Common throughout the state; in the coastal plain in oak-hickory, beech, and pine–mixed hardwood forests.

Comments: Dogwood is commonly cultivated for its "flowers." The berries are an important source of food for wildlife but are poisonous to humans. The hardness of its wood makes it excellent for shuttles in the textile industry; also used for mauls, mallets, wedges, and pulleys.

The powdered bark of dogwood was one of the most important sources of a drug during the Civil War. The drug was used as a substitute for quinine that became unavailable due to the blockade by Union forces. The twigs of dogwood have been used as chewsticks to clean teeth.

213 **Sassafras**

Sassafras albidum (Nuttall) Nees
Lauraceae (Laurel Family)
Flowers: March–April
Description: Small- to medium-sized, fast-growing tree, often forming thickets from lateral root-sprouts; male and female flowers on separate plants, appearing in the spring before the leaves; leaves deciduous, alternate, and polymorphic (either entire, with a one-sided lobe, or three-lobed); twigs and leaves spicy, aromatic.
Range-Habitat: Common throughout the state; in the coastal plain common in thin oak-hickory forests, along fence rows, abandoned fields, pine mixed hardwood forests, and woodland borders.
Comments: Sassafras is an early successional tree, and because it is shade intolerant, is seldom found as an understory tree in a closed canopy. It is one of the first trees to invade abandoned fields, often appearing as small groves spreading from lateral root shoots. Fruit production is sparse but does provide an important food for wildlife.

Oil of sassafras is distilled from the bark of the roots and is a flavoring material of considerable importance. The oil is used to flavor tobacco, patent medicines, root beer and other beverages, soaps, perfumes, and gums. A tea made from boiling the bark was once used as a spring tonic to "thin the blood" before the advent of summer; proven ineffective, it quickly fell into disrepute. Young leaves are ground into a fine powder to produce the mucilaginous gumbo of Creole cooking.

Warning: The Federal Drug Administration has banned the use of safrole, found in oil of sassafras, as possibly carcinogenic.

214 **Horse Sugar; Sweetleaf; Yellow-wood**

Symplocos tinctoria (L.) L'Her
Symplocaceae (Sweetleaf Family)
Flowers: March–May
Description: Shrub or small tree; leaves simple, alternate, deciduous, or sometimes tardily deciduous; flowers fragrant, in dense clusters on naked branches (or almost naked if leaves persist).
Range-Habitat: Primarily a mountains and coastal plain species; in the coastal plain in a wide variety of habitats: oak-hickory and beech forests, longleaf pine flatwoods, maritime forests, thickets, and stream margins.
Comments: The bark and leaves yield a bold yellow dye; dying with yellow-wood, a practice several centuries old in the South, has never been lost as an art. The leaves are greedily consumed by cattle and horses and browsed by whitetail

deer. Older leaves often show signs of insect damage. The leaves can be chewed as a refreshing trail tidbit. The wood is of no commercial value.

215 **Eastern Redbud; Judas-tree**

Cercis canadensis L.
Fabaceae (Pea or Bean Family)
Flowers: March–May
Description: Small, short-lived tree with alternate, simple, deciduous, heart-shaped leaves; flowers borne on old wood, appearing in spring before the leaves.
Range-Habitat: Common throughout the state; in the coastal plain in oak-hickory, beech, and pine–mixed hardwood forests; especially common in calcareous soil.
Comments: Redbud is one of the most popular native trees used for cultivation. Redbud is an understory tree whose trunk is too small to be commercially important. The flowers have an acidlike flavor and are put into salads; also pickled in the bud. The buds, flowers, and young pods are good fried in butter or made into fritters.

The other common name, Judas-tree, is sometimes transferred to the eastern redbud from the related species of the Mediterranean (*Cercis siliquastrum*)—the tree on which Judas hanged himself. Legend states that the flowers were white but turned to red either with shame or from the drops of blood shed by Jesus.

216 **Yellow Jessamine**

Gelsemium sempervirens (L.) Aiton f.
Loganiaceae (Logania Family)
Flowers: March–Early May
Description: Twining (left to right), woody vine with opposite, pointed, evergreen leaves; trailing or high-climbing; blades smooth, entire; flowers fragrant.
Range-Habitat: Common throughout the piedmont and coastal plain; in the coastal plain in oak-hickory, beech, pine–mixed hardwood and maritime forests, fence rows, thickets, and roadsides.
Comments: This is the state flower of South Carolina and is often cultivated. All parts are poisonous when taken internally, but not to the touch. Children have been poisoned by sucking nectar from the flowers, probably mistaking them for honeysuckle.

217 **Dwarf Iris**

Iris verna L.
Iridaceae (Iris Family)
Flowers: Late March–Early May
Description: Erect, perennial herb from densely scaly rhizomes; stems (in South Carolina) to 6" tall, unbranched, hidden by sheathing bracts; leaves essentially straight, narrow; flowers

very fragrant.

Range-Habitat: A coastal plain and piedmont species; rare in the outer coastal plain in sandy oak-hickory forests, low moist thickets, and edges of pocosins.

Comments: Accordingly to Kingsbury (1964), "Various iris have been found to contain an irritant principle in the leaves or particularly in the rootstocks which produce gastroenteritis if ingested in sufficient, relatively large, amounts."

Dwarf iris is rare in the outer coastal plain with little known about its habitat requirements. I have observed it in sandy oak-hickory forests and low, moist thickets. Others have reported seeing it along borders of pocosins.

218 Coral Honeysuckle; Woodbine

Lonicera sempervirens L.

Caprifoliaceae (Honeysuckle Family)

Flowers: March–July

Description: Trailing or high-climbing, woody vine with simple, opposite leaves; partially evergreen; twining is from left to right; last one or two pairs of leaves below the inflorescence joined at their bases and thus perfoliate; berries red, maturing July–September.

Range-Habitat: Common throughout the state; in the coastal plain in thin oak-hickory and pine-hardwood forests, thickets, and fence rows.

219 Black Cherry

Prunus serotina Ehrhart

Rosaceae (Rose Family)

Flowers: April–May

Description: Large tree reaching 80–90' tall; bark of small trees with horizontal lenticles, becoming fissured and scaly with age; leaves and inner bark contain the almond-flavored hydrocyanic acid which can readily be detected by breaking a twig and smelling the broken end; leaves alternate, simple, deciduous; fruit a juicy drupe, black when mature in July–August.

Range-Habitat: Common throughout the state; in the coastal plain in a variety of natural and disturbed habitats: oak-hickory and pine–mixed hardwood forests, fence rows, thickets, pastures, and longleaf pine flatwoods.

Comments: Black cherry is one of the most versatile native trees. Its reddish brown, close-grained wood takes a beautiful polish and is widely used to make furniture, veneers, and small wooden wares. Its use as a lumber tree today is reduced because of the lack of large trees. It is now used primarily as a specialty wood.

As an astringent, the bark is used to make cough medicines and expectorants for the treatment of sore throats. The ripe fruits were used by pioneers in the Appalachians to make a drink called "cherry bounce." The juice was pressed from the fruits and infused in brandy or rum to give it the bitter taste desired. Even today, the fruit is used to flavor liqueurs.

All parts of the plant are poisonous (except the pulp of the ripe fruit) because of the hydrocyanic acid. Children have been poisoned by sucking the twigs.

The fruits are eaten by a wide variety of wildlife.

220 **Lousewort; Wood Betony**
Pedicularis canadensis L.
Scrophulariaceae (Figwort Family)
Flowers: April–May
Description: Perennial, hairy herb from thickened, fibrous roots; maybe hemiparasite; plant 6–12" tall; lower leaves in a basal cluster, stem leaves alternate, reduced; all leaves deeply divided into toothed lobes; corolla in various color combinations of yellow, red, and purplish.
Range-Habitat: Chiefly mountains and piedmont; scattered and rare in the coastal plain in rich oak-hickory forests and roadside meadows.
Comments: The generic name comes from the Latin *pediculus*, a louse, from an early European belief that cattle feeding where these plants abounded became infected with lice.

This is one of the species of figworts believed to be a hemiparasite; however, no studies have indicated the identity of the host plant.

221 **Fringe-tree**
Chionanthus virginicus L.
Oleaceae (Olive Family)
Flowers: April–May
Description: Fast-growing, short-lived shrub or small tree, occasionally reaching 30' tall; leaves simple, opposite, deciduous, and entire.
Range-Habitat: Common throughout the state; in the coastal plain in dry oak-hickory forests and pine–mixed hardwood forests.
Comments: Fringe-tree is a widely cultivated native tree in the Lowcountry; its attraction is its airy clusters of fragrant, white flowers. The drupes are eaten by a wide variety of wildlife.

222 **Bigleaf Snowbell**
Styrax grandifolia Aiton
Styracaceae (Storax Family)
Flowers: April–May
Description: Small tree, up to 20' tall, but more commonly a large shrub; flowers fragrant; leaves with star-shaped hairs beneath.
Range-Habitat: Primarily a piedmont and coastal plain species; occasional in the coastal plain as understory species in

moist oak-hickory forests, along streambanks, and in pine–mixed hardwood forests.

Comments: This attractive small tree or shrub is often used as an ornamental. It is of little value to wildlife and too small for lumber.

223 **Crab-apple; Southern Crab-apple**

Malus angustifolia (Aiton) Michaux

Synonym: *Pyrus angustifolia* Aiton

Rosaceae (Rose Family)

Flowers: April–May

Description: Small tree or thicket-forming shrub from root sprouts; petals white or light pink; deciduous, simple, alternate leaves with margins of some leaves scalloped, toothed, or nearly entire; fruit a pome, yellowish green, very sour to taste, ripe in August–September.

Range-Habitat: Throughout the state but more common in the coastal plain; scattered throughout the coastal plain but generally not common; grows in moist soil in oak-hickory forests, woodland borders, and fence rows.

Comments: Crab-apple is often used as an ornamental because of its showy and fragrant flowers. The fruit is used to make jelly, preserves, and cider; it is an important wildlife food consumed by deer, foxes, raccoons, quail, and turkeys. The wood is not commercially important.

224 **Sparkleberry**

Vaccinium arboreum Marshall

Ericaceae (Heath Family)

Flowers: Late April–June

Description: Erect shrub or small tree to 30' tall; branches crooked, forming an irregular crown; leaves simple, alternate, shiny above, evergreen to tardily deciduous.

Range-Habitat: Common throughout the state; in the coastal plain in oak-hickory, maritime, beech, and pine–mixed hardwood forests.

Comments: The berries, often lasting through winter, are eaten by a wide variety of wildlife. The pulp is scanty but pleasant tasting and can be made into jelly or jam. The wood has little commercial value; the roots, bark, and leaves were once used to treat diarrhea and dysentery.

225 **New Jersey Tea**

Ceanothus americanus L.

Rhamnaceae (Buckthorn Family)

Flowers: May–June

Description: Low shrub to 3' tall; roots dark red, large; leaves simple, alternate, deciduous, strongly three-nerved beneath.

Range-Habitat: Common throughout the state; in the coastal plain in open oak-hickory forests, longleaf pine flatwoods, and roadsides.

Comments: The dried leaves were used during the Revolutionary War as a substitute for oriental tea but contains no caffeine and does not provide the lift of oriental tea. The root is reported to be a stimulant, a sedative, and a means of loosening phlegm. Nodules on the roots harbor bacteria that fix atmospheric nitrogen.

226 **White Milkweed**
Asclepias variegata L.
Asclepiadaceae (Milkweed Family)
Flowers: May–June
Description: Perennial herb with milky juice; stem simple, solitary, 1–3' tall; leaves broad, in two to five pairs; corolla bright white; follicles maturing July–September.
Range-Habitat: Throughout the state but primarily in the mountains and piedmont; occasional in the coastal plain in dry oak-hickory forests; sandy, dry, open woods; and woodland margins.
Comments: The following account applies to white milkweed and most other species. The seed pods, when almost mature but still solid, make a palatable vegetable comparable to okra when cooked. The shoots and young leaves are used like spinach, after the first water has removed the bitter taste from the milky juice. But be warned: the raw shoots of any milkweed may be poisonous.

227 **Indian Pink**
Spigelia marilandica L.
Loganiaceae (Logania Family)
Flowers: May–June
Description: Erect, perennial herb to 28" tall; stems with four to seven pairs of opposite, sessile leaves; flower scarlet on outside, yellow-green on inside.
Range-Habitat: Scattered throughout the state; in the coastal plain frequent in rich oak-hickory and beech forests.
Comments: The demand for Indian pink in the early 1800s for use as a vermifuge almost lead to its extinction; however, severe side effects accompanied its use so that by the early 1900s it was discontinued. The plant is once again common.

228 **Hairy Skullcap**
Scutellaria elliptica Muhl.
Lamiaceae (Mint Family)
Flowers: May–June
Description: Perennial herb with four-angled stems, forming clumps of one to three stems; stems hairy, 6–30" tall; leaves opposite, generally three to five pairs below the inflorescence.
Range-Habitat: Throughout the state; occasional in the outer coastal plain in oak-hickory forests and along roadsides.

229 **Virginia Creeper**
Parthenocissus quinquefolia (L.) Planchon
Vitaceae (Grape Family)
Flowers: May–July
Description: High-climbing, woody vine with white pith; climbs by tendrils that emerge from the stem and have adhesive disks; leaves palmately compound into five leaflets; drupes black or dark blue, maturing July–August.
Range-Habitat: Common throughout the state; occurs in a variety of natural and disturbed habitats: oak-hickory, beech, pine–mixed hardwood and maritime forests, fence rows and buildings, stable dunes, and maritime shrub thickets.
Comments: The fruits are poisonous.

230 **Butterfly-weed; Pleurisy-root; Wind Root**
Asclepias tuberosa L.
Asclepiadaceae (Milkweed Family)
Flowers: May–August
Description: Perennial herb from a thick root; stems 1–2.5' tall, one to several erect, ascending, or creeping; juice not milky; plants rough-hairy throughout; leaves abundant, alternate.
Range-Habitat: Common throughout the state; in the coastal plain in sandy oak-hickory forests; sandy, dry, open woods; dry fields; roadsides; and woodland margins.
Comments: The common name butterfly-weed comes from its brilliant flowers that attract butterflies. Pleurisy-root comes from Indians chewing the root as a cure for pleurisy and other pulmonary ailments. Wind root comes from its use to relieve flatulence. A tea made of fresh leaves has been used in Appalachia to induce vomiting. The roots of butterfly-weed are poisonous.

231 **Ruellia**
Ruellia caroliniensis (Walter) Steudel
Acanthaceae (Acanthus Family)
Flowers: May–September
Description: Perennial herb, 4–24" tall; leaves entire, opposite.
Range-Habitat: Common throughout the state; in the coastal area common in dry oak-hickory forests and sandy, dry, open woods.

232 **Flowering Spurge**
Euphorbia corollata L.
Euphorbiaceae (Spurge Family)
Flowers: May–September
Description: Erect perennial with milky juice; stem freely branched above, 10–36" tall; leaves symmetrical, alternate below, opposite in the inflorescence.

Range-Habitat: Common throughout the state; in the coastal plain in oak-hickory forests, longleaf pine flatwoods, pine–mixed hardwood forests, roadsides, fields, and along railroads.

Comments: The flower construction of *Euphorbia* is very complex. The reader may refer to standard manuals for this information.

Flowering spurge is reported as poisonous; the juice may cause skin irritation.

233 **Butterfly Pea**

Clitoria mariana L.

Fabaceae (Pea or Bean Family)

Flowers: June–August

Description: Trailing or twining, perennial, herbaceous vine; stems to 3' long, little branched; leaves trifoliolate; calyx tube much longer than the lobes; standard petal below the others. (In most legumes it is above.)

Range-Habitat: Common throughout the state; in the coastal plain in oak-hickory forests, dry flats behind coastal dunes, longleaf pine flatwoods, and sandy, dry, open woods.

234 **Devil's Walking Stick**

Aralia spinosa L.

Araliaceae (Ginseng Family)

Flowers: June–September

Description: Fast-growing shrub or short-lived tree, often spreading by underground runners, reaching heights of 25–30'; stems and leaves coarsely prickly; leaves bipinnately compound, alternate; drupes purple or black, maturing in fall.

Range-Habitat: Scattered throughout the state; common in the coastal plain in oak-hickory forests, pine–mixed hardwood forests, and pocosins.

Comments: This plant has been used in a variety of folk remedies. Indians used a decoction of bark and roots to purify the blood and to treat fever. The roots, bark, and berries have been used as stimulants, and rural people applied a dried root powder to treat snakebite.

Walking stick is often mistaken for elderberry, the berries of which are edible; the former can readily be distinguished by the prickles on the leaves and stems. Walking stick berries are suspected of poisoning livestock and should be considered poisonous to humans.

235 **Indian Pipe**

Monotropa uniflora L.

Ericaceae (Heath Family)

Flowers: June–October

Description: Saprophytic herb with no chlorophyll; stem 2–8" tall; leaves reduced to scales; specimens vary from white, pink, red, pale yellow to lavender, or a combination of two of these; flowers solitary at end of stem, drooping; capsules become erect as they mature in August–November.

Range-Habitat: Throughout the state; in the coastal plain occasional in oak-hickory and pine–mixed hardwood forests.

Comments: See "Saprophytic Plants and Mycotrophy" in the "Natural History"chapter in part 1 for a more detailed coverage of Indian pipe.

Similar Species: Pine-sap (*Monotropa hypopithys* L.) can readily be separated from Indian pipe because it has numerous flowers in racemes on the end of the stem. It is almost exclusively a mountains and piedmont species; I did find a single specimen at the Bluff Plantation Wildlife Sanctuary in Berkeley County.

236 **Elephant's-foot**

Elephantopus tomentosus L.

Asteraceae (Aster or Sunflower Family)

Flowers: Late July–September

Description: Perennial herb to 2' tall; leaves mostly basal, up to 12" long, soft-hairy beneath; flower heads with three conspicuous resin-dotted bracts below.

Range-Habitat: Common throughout the state; in the coastal plain in oak-hickory forests, longleaf pine flatwoods, roadsides, maritime forests, and pine–mixed hardwood forests.

237 **Bear's-foot; Yellow Leafcup**

Polymnia uvedalia L.

Asteraceae (Aster or Sunflower Family)

Flowers: July–October

Description: Perennial herb with thick, fleshy roots; stems erect, hollow, 3–10' tall; glandular or spreading-hairy beneath the heads; leaves opposite, palmately lobed, or cut with their shape promoting the common name—bear's-foot.

Range-Habitat: Throughout the state, but less common in the coastal plain; occasional in the coastal plain in oak-hickory forests, pastures, edge of woodland borders, and thickets.

238 **Large False Foxglove**

Aureolaria flava (L.) Farwell

Scrophulariaceae (Foxglove Family)

Flowers: August–September

Description: Hemiparasite on members of the white oak group; perennial herb to 6' tall; leaves opposite; flowers open in early morning, drop by late afternoon.

Range-Habitat: Widely scattered localities throughout the state; in the coastal plain frequent in oak-hickory forests, roadsides, and dry, longleaf pine flatwoods where members of the white oak group occur.

239 **Tickweed**

Verbesina virginica L.
Synonym: *Phaethusa virginica* (L.) Small
Asteraceae (Aster or Sunflower Family)
Flowers: August–October
Description: Perennial herb with erect, densely hairy, simple stems; stems winged, to 6–10' tall; leaves alternate, with winged leaf stalks.
Range-Habitat: Common throughout the state; in the coastal plain it occurs in oak-hickory forests, old fields, pastures, road banks, and meadows.
Similar Species: *V. occidentalis* (L.) Walter is similar but has yellow ray and disk flowers; common statewide, in the same habitats in the coastal plain as *V. virginica.*

240 **Pale Gentian**

Gentiana villosa L.
Gentianaceae (Gentian Family)
Flowers: Late August–November
Description: Perennial herb with erect stems, 6–20" tall; leaves opposite, five to twelve pairs; corolla greenish white, often tinged with purple; corolla with pleats (a thin tissue connecting the lobes).
Range-Habitat: Scattered throughout the state; in the coastal plain occasional in rich oak-hickory forests and marl forests.

241 **Chinquapin**

Castanea pumila (L.) Miller
Fabaceae (Beech Family)
Fruits: September–October
Description: Large shrub or small tree; trunk one to several, soon branching; leaves alternate, deciduous, with numerous large teeth along the margin; leaves whitish and hairy below; male and female flowers on different catkins, or sometimes female flowers at base of a catkin with male flowers above; nut surrounded by spiny bur.
Range-Habitat: Common throughout the state; in the coastal plain in dry oak-hickory forests or open, pine–mixed hardwood forests.
Taxonomy: Two varieties are recognized: var. *pumila* with a dense spine cluster on the fruits and var. *ashei* with spine clusters distant. Both occur on the coastal plain, but the latter is confined to the outer coastal plain in dry, sandy soils.
Comments: The nuts are sweet and can be used in the same ways the American chestnut (*Castanea dentata*) was used: eaten raw, boiled with salt, boiled and mashed like potatoes, or boiled and dipped in sugar.

242 **Mockernut Hickory**

Carya tomentosa (Poiret) Nuttall
Juglandaceae (Walnut Family)
Fruits: October
Description: Large tree to 90' tall; winter buds with dark, reddish brown, hairy scales; leaves alternate, odd-pinnately compound with five to nine leaflets; twigs bend but do not break; clusters of hairs present on lower surface of leaflets.
Range-Habitat: Common throughout the state; in the coastal plain in oak-hickory and beech forests, longleaf pine flatwoods, and sandy, dry, open woods.
Comments: Mockernut is the most common hickory in the southern United States. The common name comes from the thick shells that enclose the sweet kernels, making it difficult to remove the meat; hence, the name mock meaning to defy. Native Americans beat these nuts into pieces with stones, put the pieces (shells included) into mortars, added water, and pounded until a kind of milk called "pochoicora" was produced. Hickory comes from this Indian word.

The wood is tough, hard, and strong with good bending qualities. It is used for tool handles, wood splints, and rustic furniture.

243 *Sandy, Dry, Open Woods*

244 **Blue Star**

Amsonia ciliata Walter
Apocynaceae (Dogbane Family)
Flowers: April
Description: Perennial herb with hairy stems; 12–28" tall; leaves alternate, filiform to linear; the star-shaped corolla is distinctive.
Range-Habitat: Primarily from the inner coastal plain; known only in Jasper County in the outer coastal plain; found in sandy, dry, open woods and xeric sandhills.
Comments: Blue star occurs at the Tillman Sand Ridge Heritage Preserve and environs in Jasper County.

245 **Sundial Lupine**

Lupinus perennis L.
Fabaceae (Pea or Bean Family)
Flowers: April–May
Description: Perennial, erect herb 8–24" tall; from creeping underground rhizome; leaves palmately compound with seven to eleven leaflets.
Range-Habitat: A coastal plain species; frequent in sandy, dry, open woods and clearings.
Comments: The genus name *Lupinus* comes from the Latin *lupus* (wolf) because they were thought to deplete or "wolf"

the soil of minerals. We now know this is unfounded and that members of the pea family actually enrich the soil by harboring bacteria in their roots that fix atmospheric nitrogen into a useful form.

246 **Lady Lupine**
Lupinus villosus Willd.
Fabaceae (Pea or Bean Family)
Flowers: April–May
Description: Perennial herb with creeping stems from a deep, woody taproot; stems generally in dense clumps, shaggy-hairy throughout; leaves evergreen and simple; flowering stalks to 20" high; standard with deep reddish purple spot; legume shaggy-hairy, maturing June–August.
Range-Habitat: Common in the coastal plain in sandy, dry, open woods and sandy roadsides.
Similar Species: *L. diffusus* Nuttall is similar, but the standard has a conspicuous white to cream spot. It is primarily a coastal plain species; occasional in the outer coastal plain in sandy, dry, open woods and sandy roadsides; flowers in March–May.

247 **Deerberry; Gooseberry; Squaw-huckleberry**
Vaccinium stamineum L.
Ericaceae (Heath Family)
Flowers: April–June
Description: Deciduous shrub to 16' tall (or rarely found taller); flowers or fruits with small, leafy bracts; fruits bitter, green, pink, yellowish, or purple, ripe August–October.
Range-Habitat: Common throughout the state; in the coastal plain in sandy, dry, open woods; thin, live oak woods; xeric sandhills; and longleaf pine flatwoods.
Comments: The specific epithet *stamineum* means "with prominent stamens" in reference to the exerted stamens. The common name indicates the fruits are eaten by deer (and other animals). The berries are inedible when raw but are sometimes used by mountain people to make pies.

248 **Wild Pink; Carolina Pink**
Silene caroliniana Walter
Caryophyllaceae (Pink Family)
Flowers: April–July
Description: Tufted perennial up to 8" tall from a thin, deep taproot; leaves oblong, well separated, up to 2" long; petals vary from pink to white, about .5" long.
Range-Habitat: A lower piedmont and coastal plain species; occasional in sandy, dry, open woods.

249 **Prostrate Rattlebox; Rabbit-bells**
Crotalaria angulata Miller
Synonym: *Crotalaria rotundifolia* (Walter) Poiret
Fabaceae (Pea or Bean Family)

Flowers: April–August
Description: Perennial herb to 16" tall; stems numerous, creeping to ascending, hairy, and loosely spreading; leaves alternate, simple; legume inflated, maturing in June–October.
Range-Habitat: A coastal plain and piedmont species; common in the coastal plain in sandy, dry, open woods and sandy roadsides.
Similar Species: *C. purshii* DC. has closely appressed hairs and smooth upper leaf surface; flowers July–September; habitat and range similar.
Comments: The seeds are poisonous.

250 **Queen's-delight**
Stillingia sylvatica Garden
Euphorbiaceae (Spurge Family)
Flowers: May–July
Description: Perennial herb, usually with several stems from a rootstock; to 32" tall; leaves alternate, simple; male and female flowers on terminal spike with male uppermost, female at base; spike rachis with numerous, large glands; capsule three-loculate, one seed per locule.
Range-Habitat: A coastal plain species; common in sandy, oak-hickory woods; longleaf pine flatwoods; and open pine–mixed hardwood forests.
Comments: Queen's-delight is a component in S.S.S. Tonic sold today. It has been used for a variety of home remedies. In the Old South it was used to treat constipation and to induce vomiting. Indian women drank the mashed roots boiled in water after giving birth.

251 **Roseling**
Tradescantia rosea Vent.
Commelinaceae (Dayflower Family)
Flowers: May–July
Description: Tufted, perennial herb, 8–20" tall; plant smooth or nearly so; leaves linear, sheaths usually fringed.
Range-Habitat: Chiefly coastal plain and piedmont; in the coastal plain common in sandy, dry, open woods.
Taxonomy: Two varieties are recognized: var. *rosea* with leaves more than .12" wide and var. *graminea* with leaves less than .12" wide. Var. *rosea* occurs in the sandy, dry, open woods and grows to around 8" high.

252 **Piriqueta**
Piriqueta caroliniana (Walter) Urb.
Turneraceae (Turnera Family)
Flowers: May–September
Description: Perennial herb to 20" tall; spreading and forming colonies from root sprouts; flowers remain open only on sunny days, and the petals fall off easily.

Range-Habitat: Occasional in sandy, dry, open woods in the western coastal plain counties.

253 **Hairy False Foxglove**
Aureolaria pectinata (Nuttall) Penn.
Synonym: *Gerardia pectinata* (Nuttall) Benth.
Scrophulariaceae (Figwort Family)
Flowers: May–September
Description: Densely, glandular-hairy, annual herb; to 3' tall; leaves deeply dissected.
Range-Habitat: Primarily piedmont and inner coastal plain; occasional in the outer coastal plain where it occurs in sandy, dry, open woods and thin, dry, oak-hickory forests.
Comments: Hemiparasite on the roots of *Quercus* ssp.

254 **Stylisma**
Bonamia patens (Desr.) Shinners
Synonym: *Stylisma patens* (Desr.) Myint
Convolvulaceae (Morning-glory Family)
Flowers: June–August
Description: Prostrate or spreading herbaceous, perennial vine; stems several to many, to 6' long; leaves alternate, hairy, elliptic; flowers solitary on axillary stalks.
Range-Habitat: A coastal plain species, but chiefly inner coastal plain; in the outer coastal plain occasional in sandy, dry, open woods and sandy, open roadsides.

255 **Cottonweed**
Froelichia floridana (Nuttall) Moq.
Amaranthaceae (Amaranth Family)
Flowers: June–October
Description: Stiffly, erect herb, 3–6' tall, usually with a few well-developed branches; branching mostly from upper nodes; stems and leaves with whitish hairs; leaves opposite, entire, few.
Range-Habitat: A coastal plain species; common in sandy, dry, open woods; sandy fields; and sandy roadsides.
Similar Species: *F. gracilis* (Hooker) Moq. branches mostly from the base, is usually 28" tall or less, and its fresh spikes are gray. Primarily an inner coastal plain species; occasional in sandy, dry, open woods; sandy roadsides; sandy fields; and along railroads.

256 **Dayflower**
Commelina erecta L.
Commelinaceae (Spiderwort Family)
Flowers: June–October
Description: Perennial from thickened, black roots; erect or creeping, 6–24" tall; two petals blue, the third white and smaller, generally all three wilting before noon; flowers enclosed by a large bract.

Plate 60. Needle Rush (*Juncus roemerianus*)
Flowers: May–October

Plate 61. Soft-stem Bulrush (*Scirpus validus*)
Flowers: June–September

Plate 62. Leafy Three-square (*Scirpus robustus*)
Flowers: July–September

Plate 63. Arrow-leaf Morning-glory (*Ipomoea sagittata*) Flowers: July–September

Plate 64. Saw-grass (*Cladium jamaicense*) Flowers: July–October

Plate 65. Tidal Freshwater Marsh

Plate 66. Native Wisteria (*Wisteria frutescens*)
Flowers: April–May

Plate 67. Indigo-bush (*Amorpha fruticosa*)
Flowers: April–June

Plate 68. Alligator-weed (*Alternanthera philoxeroides*)
Flowers: April–October

Plate 69. Water-spider Orchid (*Habenaria repens*)
Flowers: April–Frost

Plate 70. Spider-lily (*Hymenocallis crassifolia*)
Flowers: Mid-May–June

Plate 71. Arrow Arum (*Peltandra virginica*)
Flowers: May–June

Plate 72. Common Cattail (*Typha latifolia*)
Flowers: May–July

Plate 73. Swamp Rose (*Rosa palustris*)
Flowers: May–July

Plate 74. Obedient Plant (*Physostegia leptophylla*)
Flowers: Late May–July

Plate 75. Water Hemlock (*Cicuta maculata*)
Flowers: May–August

Plate 76. Water Primrose (*Ludwigia uruguayensis*)
Flowers: May–September

Plate 77. Pickerelweed (*Pontederia cordata*)
Flowers: May–October

Plate 78. Wapato (*Sagittaria lancifolia*) Flowers: May–November

Plate 79. Jewelweed (*Impatiens capensis*) Flowers: May–Frost

Plate 80. Ground-nut (*Apios americana*) Flowers: June–August

Plate 81. Large Marsh-pink (*Sabatia dodecandra*)
Flowers: June–August

Plate 82. Water Parsnip (*Sium suave*)
Flowers: June–August

Plate 83. Swamp Rose Mallow (*Hibiscus moscheutos*)
Flowers: June–September

Plate 84. Water Willow (*Decodon verticillatus*)
Flowers: July–September

Plate 85. Eryngo (*Eryngium aquaticum*)
Flowers: July–September

Plate 86. Hairy Hydrolea (*Hydrolea quadrivalvis*)
Flowers: July–September

Plate 87. Marsh Bulrush (*Scirpus cyperinus*)
Flowers: July–September

Plate 88. Seashore Mallow (*Kosteletskya virginica*)
Flowers: July–October

Plate 89. Climbing Hempweed (*Mikania scandens*)
Flowers: July–October

Plate 90. Cardinal Flower (*Lobelia cardinalis*)
Flowers: July–October

Plate 91. Fragrant Ladies' Tresses (*Spiranthes odorata*)
Flowers: July–Frost

Plate 92. Dodder (*Cuscuta gronovii*)
Flowers: August–October

Plate 93. Giant Beard Grass (*Erianthus giganteus*)
Flowers: September–October

Plate 94. Climbing Aster (*Aster carolinianus*)
Flowers: Late September–October

Plate 95. Bur-marigold (*Bidens laevis*)
Flowers: Late September–November

Plate 96. Tidal Freshwater Swamp Forest

Plate 97. Stream Bank

Plate 98. River Birch (*Betula nigra*)
Trunk: All Year

Plate 99. Tag Alder (*Alnus serrulata*) Flowers: February–March

Plate 100. Button-bush (*Cephalanthus occidentalis*) Flowers: June–August

Plate 101. Water Ash (*Fraxinus caroliniana*) Fruits: July–October

Plate 102. American Sycamore (*Platanus occidentalis*)
Fruits: October

The Freshwater Aquatic Gardens

Plate 103. Freshwater Aquatics

Plate 104. Duckmeat (*Spirodela polyrhiza*)
Vegetative: All Year

Bog-mat (*Wolffiella floridana*)
Vegetative: All Year

Plate 105. Floating-heart (*Nymphoides aquatica*)
Flowers: Late April–September

Plate 106. Cow-lily (*Nuphar luteum*)
Flowers: April–October

Plate 107. Purple Bladderwort (*Utricularia purpurea*)
Flowers: May–September

Plate 108. Floating Bladderwort (*Utricularia inflata* var. *inflata*)
Flowers: May–November

Plate 109. Water Hyacinth (*Eichhornia crassipes*)
Flowers: June–September

Plate 110. Sacred Bean (*Nelumbo lutea*)
Flowers: June–September

Plate 111. Fragrant Water-lily (*Nymphaea odorata*)
Flowers: June–September

Plate 112. Pondweed (*Potamogeton plucher*)
Flowers: June–September

Plate 113. Frog's-bit (*Limnobium spongea*)
Flowers: June–September

Plate 114. Water-shield (*Brasenia schreberi*)
Flowers: June–October

The Peatland Gardens

Plate 115. Pocosin

Plate 116. Pond Pine (*Pinus serotina*)
Female Cones: All Year

Plate 117. Highbush Blueberry (*Vaccinium corymbosum*)
Flowers: Late February–May

Plate 118. Leather-leaf (*Cassandra calyculata*)
Flowers: March–April

Plate 119. Creeping Blueberry (*Vaccinium crassifolium*)
Flowers: April–May

Plate 120. Witch-alder (*Fothergilla gardenii*)
Flowers: April–May

Plate 121. Possum-haw (*Viburnum nudum*)
Flowers: April–May

Plate 122. Sweet Pitcher-plant (*Sarracenia rubra*)
Flowers: April–May

Plate 123. Frog's Britches (*Sarracenia purpurea*)
Flowers: April–May

Plate 124. Lamb-kill (*Kalmia angustifolia* var. *caroliniana*)
Flowers: April–Early June

Plate 125. Fetterbush (*Lyonia lucida*)
Flowers: April–Early June

Plate 126. Honeycup (*Zenobia pulverulenta*)
Flowers: April–June

Plate 127. Sweet Bay (*Magnolia virginica*) Flowers: April–July

Plate 128. Poison Sumac (*Rhus vernix*) Flowers: May–Early June

Plate 129. Rose Pogonia (*Pogonia ophioglossoides*) Flowers: May–June

Plate 130. Swamp Honeysuckle (*Rhododendron viscosum*)
Flowers: May–July

Plate 131. Titi (*Cyrilla racemiflora*)
Flowers: May–July

Plate 132. White Arum (*Peltandra sagittaefolia*)
Flowers: June–August

Plate 133. Loblolly Bay (*Gordonia lasianthus*) Flowers: July–September

Plate 134. Bamboo-vine (*Smilax laurifolia*) Fruits: September–October

The Bottomland Forest Gardens

Plate 135. Bald Cypress–Tupelo Gum Swamp Forest

Plate 136. Mistletoe (*Phoradendron serotinum*)

Plate 137. Bald Cypress (*Taxodium distichum*)
Female Cone: All Year

Plate 138. Golden-club (*Orontium aquaticum*)
Flowers: March–April

Plate 139. Red Maple
(*Acer rubrum*)
Fruits: March–May

Plate 140. Butterweed
(*Senecio glabellus*)
Flowers: March–June

Plate 141. Leucothoe
(*Leucothoe racemosa*)
Flowers: March–Early June

Plate 142. Supplejack (*Berchemia scandens*)
Flowers: April–May

Plate 143. Cross Vine (*Anisostichus capreolata*)
Flowers: April–May

Plate 144. Blue Flag Iris (*Iris virginica*)
Flowers: April–May

Plate 145. Swamp Dogwood (*Cornus stricta*)
Flowers: April–May

Plate 146. Snowbell (*Styrax americana*)
Flowers: April–June

Plate 147. Virginia Willow (*Itea virginica*)
Flowers: May–June

Plate 148. Water Willow (*Justicia ovata*)
Flowers: May–July

Plate 149. Lizard's Tail (*Saururus cernuus*)
Flowers: May–July

Plate 150. Bur-reed (*Sparganium americanum*)
Flowers: May–September

Plate 151. Ladies'-eardrops (*Brunnichia cirrhosa*)
Flowers: June–July

Plate 152. Aquatic Milkweed (*Asclepias perennis*)
Flowers: June–August

Plate 153. Green-fly Orchid (*Epidendrum conopseum*)
Flowers: July–September

Plate 154. Macbridea (*Macbridea caroliniana*) Flowers: July–August

Plate 155. Elderberry (*Sambucus canadensis*) Fruits: July–August

Plate 156. False Nettle (*Boehmeria cylindrica*) Flowers: July–August

Plate 157. Skullcap (*Scutellaria lateriflora*)
Flowers: July–Frost

Plate 158. Purple Lobelia (*Lobelia elongata*)
Flowers: August–October

Plate 159. Poison Ivy (*Rhus radicans*)
Fruits: August–October

Plate 160. Tupelo Gum (*Nyssa aquatica*)
Fruits: September–October

Plate 161. Coral Greenbrier (*Smilax walterii*)
Fruits: September–Winter

Plate 162. Hardwood Bottom

Plate 163. Jack-in-the-pulpit (*Arisaema triphyllum*)
Flowers: March–April

Plate 164. Pawpaw (*Asimina triloba*)
Flowers: March–May

Plate 165. Easter Lily (*Zephyranthes atamasco*)
Flowers: Late March–April

Plate 166. Dog-hobble (*Leucothoe axillaris*) Flowers: Late March–May

Plate 167. Arrowwood (*Viburnum dentatum*) Flowers: Late March–May

Plate 168. Leather-flower (*Clematis crispa*)
Flowers: April–August

Plate 169. Fever Tree (*Pinckneya pubens*)
Flowers: May

Plate 170. Climbing Hydrangea (*Decumaria barbara*)
Flowers: May–June

Plate 171. Cow-itch (*Campsis radicans*)
Flowers: June–July

Plate 172. River Oats (*Uniola latifolia*)
Flowers: June–October

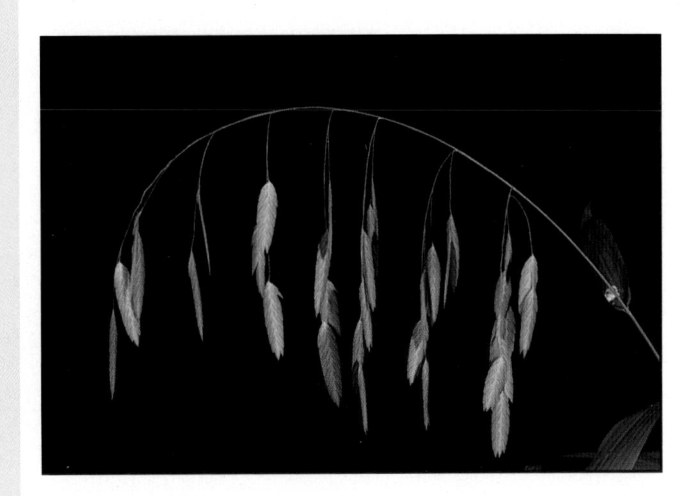

Plate 173. Sweet-gum (*Liquidambar styraciflua*)
Fruits: Fall–Winter

The Deciduous Forest Gardens

Plate 174. Marl Forest

Plate 175. Blackstem Spleenwort (*Asplenium resiliens*)

Plate 176. Meadow Parsnip (*Thaspium barbinode*)
Flowers: April–May

Plate 177. Alumroot
(*Heuchera americana*)
Flowers: April–June

Plate 178. Thimbleweed
(*Anemone virginiana*)
Flowers: May–July

Plate 179. Elytraria (*Elytraria caroliniensis*)
Flowers: June–August

Plate 180. Crested Coral-root (*Hexalectris spicata*)
Flowers: July–August

Plate 181. White Basswood (*Tilia heterophylla*)
Fruits: July–August

Plate 182. Horse-balm (*Collinsonia tuberosa*)
Flowers: Late July–September

Plate 183. Hop Hornbeam (*Ostrya virginiana*)
Fruits: August–October

Plate 184. Buckthorn (*Rhamnus caroliniana*)
Fruits: September–October

Plate 185. Beech Forest

Plate 186. Bloodroot (*Sanguinaria canadensis*) Flowers: Early March–April

Plate 187. Blue Star (*Amsonia tabernaemontana*) Flowers: March–April

Plate 188. Pennywort (*Obolaria virginica*)
Flowers: March–April

Plate 189. May-apple (*Podophyllum peltatum*)
Flowers: March–April

Plate 190. Little Sweet Betsy (*Trillium cuneatum*)
Flowers: Mid-March–April

Plate 191. Heart-leaf (*Hexastylis arifolia*)
Flowers: March–May

Plate 192. Green-and-gold (*Chrysogonum virginianum*)
Flowers: Late March–Early June

Plate 193. Cancer-root (*Conopholis americana*)
Flowers: March–June

Plate 194. Southern Twayblade (*Listera australis*)
Flowers: March–July

Plate 195. Red Buckeye (*Aesculus pavia*)
Flowers: April–Early May

Plate 196. Violet Wood Sorrel (*Oxalis violacea*) Flowers: April–May; Fall

Plate 197. Wild Ginger (*Asarum canadense*) Flowers: April–May

Plate 198. Sweet Shrub (*Calycanthus floridus*) Flowers: April–May

Plate 199. Bellwort
(*Uvularia perfoliata*)
Flowers: April–May

Plate 200. Tulip Tree
(*Liriodendron tulipifera*)
Flowers: April–June

Plate 201. Indian Cucumber-root (*Medeola virginiana*)
Flowers: April–June

Plate 202. Wild Geranium (*Geranium maculatum*)
Flowers: April–June

Plate 203. Spotted Wintergreen (*Chimaphila maculata*)
Flowers: May–June

Plate 204. Partridge Berry (*Mitchella repens*)
Flowers: May–June
Fruits: June–Winter

Plate 205. Silky Camellia (*Stewartia malacodendron*)
Flowers: May–June

Plate 206. Lopseed (*Phryma leptostachya*)
Flowers: June–August
Fruits: August–October

Plate 207. Green Adder's Mouth (*Malaxis unifolia*)
Flowers: June–August

Plate 208. American Beech (*Fagus grandifolia*)
Fruits: September–October

Plate 209. Beech-drops (*Epifagus virginiana*)
Flowers: September–November

Plate 210. Witch-hazel (*Hamamelis virginiana*)
Flowers: October–December

Plate 211. Oak-Hickory Forest

Plate 212. Flowering Dogwood (*Cornus florida*)
Flowers: March–April

Plate 213. Sassafras (*Sassafras albidum*)
Flowers: March–April

Plate 214. Horse Sugar (*Symplocos tinctoria*)
Flowers: March–May

Plate 215. Eastern Redbud (*Cercis canadensis*)
Flowers: March–May

Plate 216. Yellow Jessamine (*Gelsemium sempervirens*)
Flowers: March–Early May

Plate 217. Dwarf Iris (*Iris verna*)
Flowers: Late March–Early May

Plate 218. Coral Honeysuckle (*Lonicera sempervirens*)
Flowers: March–July

Plate 219. Black Cherry (*Prunus serotina*)
Flowers: April–May

Plate 220. Lousewort (*Pedicularis canadensis*)
Flowers: April–May

Plate 221. Fringe-tree (*Chionanthus virginicus*)
Flowers: April–May

Plate 222. Bigleaf Snowbell (*Styrax grandifolia*)
Flowers: April–May

Plate 223. Crab-apple (*Malus angustifolia*)
Flowers: April–May

Plate 224. Sparkleberry (*Vaccinium arboreum*)
Flowers: Late April–June

Plate 225. New Jersey Tea (*Ceanothus americanus*)
Flowers: May–June

Plate 226. White Milkweed (*Asclepias variegata*)
Flowers: May–June

Plate 227. Indian Pink (*Spigelia marilandica*)
Flowers: May–June

Plate 228. Hairy Skullcap (*Scutellaria elliptica*)
Flowers: May–June

Plate 229. Virginia Creeper (*Parthenocissus quinquefolia*)
Flowers: May–July

Plate 230. Butterfly-weed (*Asclepias tuberosa*)
Flowers: May–August

Plate 231. Ruellia (*Ruellia caroliniensis*)
Flowers: May–September

Plate 232. Flowering Spurge (*Euphorbia corollata*)
Flowers: May–September

Plate 233. Butterfly Pea (*Clitoria mariana*)
Flowers: June–August

Plate 234. Devil's Walking Stick (*Aralia spinosa*)
Flowers: June–September

Plate 235. Indian Pipe (*Monotropa uniflora*)
Flowers: June–October

Plate 236. Elephant's-foot (*Elephantopus tomentosus*)
Flowers: Late July–September

Plate 237. Bear's-foot (*Polymnia uvedalia*)
Flowers: July–October

Plate 238. Large False Foxglove (*Aureolaria flava*)
Flowers: August–September

Plate 239. Tickweed (*Verbesina virginica*)
Flowers: August–October

Plate 240. Pale Gentian (*Gentiana villosa*)
Flowers: Late August–November

Plate 241. Chinquapin (*Castanea pumila*)
Fruits: September–October

Plate 242. Mockernut Hickory (*Carya tomentosa*)
Fruits: October

Plate 243. Sandy, Dry, Open Woods

Plate 244. Blue Star
(*Amsonia ciliata*)
Flowers: April

Plate 245. Sundial Lupine (*Lupinus perennis*)
Flowers: April–May

Plate 246. Lady Lupine (*Lupinus villosus*)
Flowers: April–May

Plate 247. Deerberry (*Vaccinium stamineum*)
Flowers: April–June

Plate 248. Wild Pink (*Silene caroliniana*)
Flowers: April–July

Plate 249. Prostrate Rattlebox (*Crotalaria angulata*)
Flowers: April–August

Plate 250. Queen's-delight (*Stillingia sylvatica*)
Flowers: May–July

Plate 251. Roseling (*Tradescantia rosea*)
Flowers: May–July

Plate 252. Piriqueta (*Piriqueta caroliniana*)
Flowers: May–September

Plate 253. Hairy False Foxglove (*Aureolaria pectinata*)
Flowers: May–September

Plate 254. Stylisma
(*Bonamia patens*)
Flowers: June–August

Plate 255. Cottonweed
(*Froelichia floridana*)
Flowers: June–October

Plate 256. Dayflower
(*Commelina erecta*)
Flowers: June–October

Plate 257. Horse Mint (*Monarda punctata*)
Flowers: July–September

Plate 258. Silver-leaved Grass (*Heterotheca graminifolia*)
Flowers: July–October

Plate 259. Prickly-pear (*Opuntia compressa*)
Flowers: August–October

Plate 260. Blue Curls (*Trichostema dichotomum*)
Flowers: August–Frost

Plate 261. Blazing-star (*Liatris elegans*)
Flowers: September–October

The Pineland Gardens

Plate 262. Xeric Sandhill

Plate 263. Longleaf Pine (*Pinus palustris*)
Female Cones: All Year

Plate 264. Reindeer Moss (*Cladonia evansii*)

Plate 265. Carolina Ipecac (*Euphorbia ipecacuanhae*)
Flowers: March–May

Plate 266. Dwarf Huckleberry (*Gaylussacia dumosa*)
Flowers: March–June

Plate 267. Tread-softly (*Cnidoscolus stimulosus*)
Flowers: Late March–August

Plate 268. Gopherweed (*Baptisia perfoliata*)
Flowers: April–May

Plate 269. Gopher Apple (*Chrysobalanus oblongifolius*)
Flowers: May–June

Plate 270. Sandhills Baptisia (*Baptisia cinerea*)
Flowers: May–June

Plate 271. Sandhills Milkweed (*Asclepias humistrata*)
Flowers: May–June

Plate 272. Beard-tongue (*Penstemon australis*)
Flowers: May–July

Plate 273. Thistle (*Carduus repandus*)
Flowers: May–July

Plate 274. Warea (*Warea cuneifolia*)
Flowers: July–September

Plate 275. Soft-haired Coneflower (*Rudbeckia mollis*)
Flowers: Late August–October

Plate 276. Summer-farewell (*Petalostemum pinnatum*)
Flowers: August–Frost

Plate 277. Turkey Oak (*Quercus laevis*)
Fruits: September–October

Plate 278. Woolly Golden-aster (*Heterotheca gossypina*)
Flowers: September–October

Plate 279. Gerardia (*Agalinis setacea*)
Flowers: September–October

Plate 280. Rose Dicerandra (*Dicerandra odoratissima*)
Flowers: September–October

Plate 281. Rosemary (*Ceratiola ericoides*)
Flowers: October–November

Plate 282. Longleaf Pine Flatwood

Plate 283. Cinnamon Fern (*Osmunda cinnamomea*)
Fruits: March–May

Plate 284. Red Chokeberry (*Sorbus arbutifolia*)
Flowers: March–May

Plate 285. Dwarf Azalea (*Rhododendron atlanticum*)
Flowers: April–May

Plate 286. Stagger-bush (*Lyonia mariana*)
Flowers: April–May

Plate 287. White-topped Aster (*Aster reticulatus*)
Flowers: Late April–Early June

Plate 288. Male-berry (*Lyonia ligustrina*)
Flowers: April–June

Plate 289. Sundrops (*Oenothera fruticosa*)
Flowers: April–August

Plate 290. Yellow False-indigo (*Baptisia tinctoria*)
Flowers: April–August

Plate 291. American Chaff-seed (*Schwalbea americana*)
Flowers: May–June

Plate 292. Black-root (*Pterocaulon pycnostachyum*)
Flowers: May–June

Plate 293. Goat's Rue (*Tephrosia virginiana*)
Flowers: May–June

Plate 294. Phlox (*Phlox carolina*)
Flowers: May–July

Plate 295. White Wild Indigo (*Baptisia alba*)
Flowers: May–July

Plate 296. Curly Milkweed (*Asclepias amplexicaulis*)
Flowers: May–July

Plate 297. Sweet Pepperbush (*Clethra alnifolia*)
Flowers: May–July

Plate 298. Lead Plant (*Amorpha herbacea*)
Flowers: May–July

Plate 299. Sneezeweed (*Helenium flexuosum*)
Flowers: May–August

Plate 300. Rattlesnake Master (*Eryngium yuccifolium*)
Flowers: June–August

Plate 301. Pencil Flower (*Stylosanthes biflora*)
Flowers: June–August

Plate 302. Ironweed (*Vernonia acaulis*)
Flowers: Late June–August

Plate 303. Whorled-leaf Coreopsis (*Coreopsis major*)
Flowers: June–August

Plate 304. Blazing-star (*Liatris squarrosa*)
Flowers: June–August

Plate 305. Bicolor (*Lespedeza bicolor*)
Flowers: June–September

Plate 306. Spiked Medusa (*Eulophia ecristata*)
Flowers: June–September

Plate 307. Dollar-weed (*Rhynchosia reniformis*) Flowers: June–September

Plate 308. Sensitive Brier (*Schrankia microphylla*) Flowers: June–September

Plate 309. St. Peter's-wort (*Hypericum stans*) Flowers: June–October

Plate 310. Golden-aster (*Heterotheca mariana*) Flowers: Late June–October

Plate 311. Galactia (*Galactia macreei*) Flowers: July–September

Plate 312. Broad-leaved Eupatorium (*Eupatorium rotundifolium*) Flowers: August–October

Plate 313. Deer's-tongue (*Trilisa paniculata*) Flowers: August–October

Plate 314. Thistle (*Carduus virginianus*)
Flowers: August–Frost

Plate 315. Stiff-leaved Aster (*Aster linariifolius*)
Flowers: Late September–November

Plate 316. Toothache Grass Longleaf Pine Savanna

Plate 317. Wiregrass Longleaf Pine Savanna

Plate 318. Pixie-moss (*Pyxidanthera barbulata*)
Flowers: March–April

Plate 319. Sun-bonnets (*Chaptalia tomentosa*)
Flowers: March–May

Plate 320. Bearded Grass-pink (*Calopogon barbatus*)
Flowers: March–May

Plate 321. Yellow Butterwort (*Pinguicula lutea*)
Flowers: March–May

Plate 322. Lance-leaved Violet (*Viola lanceolata*)
Flowers: March–April

Plate 323. Leopard's-bane (*Arnica acaulis*)
Flowers: Late March–Early June

Plate 324. Hooded Pitcher-plant (*Sarracenia minor*)
Flowers: April–May

Plate 325. Indian Paint Brush (*Castilleja coccinea*)
Flowers: April–May

Plate 326. Violet Butterwort (*Pinguicula caerulea*)
Flowers: April–May

Plate 327. Sundew (*Drosera leucantha*)
Flowers: April–May

Plate 328. Black Snakeroot (*Zigadenus densus*)
Flowers: April–Early June

Plate 329. Colicroot (*Aletris farinosa*)
Flowers: Late April–Early June

Plate 330. Yellow Meadow-beauty (*Rhexia lutea*)
Flowers: April–July

Plate 331. Common Grass-pink (*Calopogon tuberosus*)
Flowers: April–July

Plate 332. Blue-hearts (*Buchnera floridana*)
Flowers: April–October

Plate 333. Orange Milkwort (*Polygala lutea*)
Flowers: April–October

Plate 334. Venus' Fly Trap (*Dionaea muscipula*)
Flowers: May–June

Plate 335. Venus' Fly Trap (*Dionaea muscipula*)

Plate 336. Milkweed (*Asclepias longifolia*)
Flowers: May–June

Plate 337. Spreading Pogonia (*Cleistes divaricata*)
Flowers: May–July

Plate 338. Narrow-leaved Skullcap (*Scutellaria integrifolia*)
Flowers: May–July

Plate 339. Samson Snakeroot (*Psoralea psoralioides*)
Flowers: May–July

Plate 340. Colicroot (*Aletris aurea*)
Flowers: Mid-May–July

Plate 341. Snowy Orchid (*Habenaria nivea*)
Flowers: May–September

Plate 342. White Sabatia (*Sabatia difformis*)
Flowers: May–September

Plate 343. Smooth Meadow-beauty (*Rhexia alifanus*)
Flowers: May–September

Plate 344. Pale Meadow-beauty (*Rhexia mariana*)
Flowers: May–October

Plate 345. Virginia Meadow-beauty (*Rhexia virginica*)
Flowers: May–October

Plate 346. Stylisma (*Bonamia aquatica*)
Flowers: June–July

Plate 347. False Asphodel (*Tofieldia racemosa*)
Flowers: June–Early August

Plate 348. Savanna Sabatia (*Sabatia campanulata*)
Flowers: June–August

Plate 349. Toothache Grass (*Ctenium aromaticum*)
Flowers: June–August

Plate 350. Dwarf Milkwort (*Polygala racemosa*)
Flowers: June–August

Plate 351. Marsh-fleabane (*Pluchea rosea*)
Flowers: June–August

Plate 352. Pineland Hibiscus (*Hibiscus aculeatus*)
Flowers: June–August

Plate 353. Redroot (*Lachnanthes caroliniana*)
Flowers: June–Early September

Plate 354. Slender Seedbox (*Ludwigia virgata*) Flowers: June–September

Plate 355. Crested Fringed-orchid (*Habenaria cristata*) Flowers: June–September

Plate 356. Bitter Mint (*Hyptis alata*) Flowers: June–September

Plate 357. Pine Lily (*Lilium catesbaei*) Flowers: Mid-June–Mid-September

Plate 358. Camass (*Zigadenus glaberrimus*)
Flowers: Late June–Early September

Plate 359. *Polygala mariana*
Flowers: June–October

Plate 360. Drum-heads (*Polygala cruciata*)
Flowers: June–October

Plate 361. Yellow Fringeless-orchid (*Habenaria integra*) Flowers: July–September

Plate 362. White Fringed-orchid (*Habenaria blephariglottis*) Flowers: July–September

Plate 363. Yellow Fringed-orchid (*Habenaria ciliaris*) Flowers: July–September

Plate 364. Blazing Star (*Liatris spicata* var. *resinosa*)
Flowers: July–September

Plate 365. Barbara's-buttons (*Marshallia graminifolia*)
Flowers: Late July–September

Plate 366. Eryngo (*Eryngium integrifolium*)
Flowers: August–October

Plate 367. Pine–Saw Palmetto Flatwood

Saw Palmetto (*Serenoa repens*)

Plate 368. Hairy Wicky (*Kalmia hirsuta*)
Flowers: June–July

Plate 369. Pine–Mixed Hardwood Forest

The Pond Cypress Gardens

Plate 370. Pond Cypress Savanna

Plate 371. Pond Cypress (*Taxodium ascendens*) Female Cones: All Year

Plate 372. Yellow Trumpet Pitcher-plant (*Sarracenia flava*) Flowers: April–May

Plate 373. Yellow Trumpet Pitcher-plant (*Sarracenia flava*)

Plate 374. Sneezeweed (*Helenium pinnatifidum*) Flowers: April–May

Plate 375. Bay Blue-flag (*Iris tridentata*)
Flowers: Late May–June

Plate 376. Tall Milkwort (*Polygala cymosa*)
Flowers: May–July

Plate 377. Sand Weed (*Hypericum fasciculatum*)
Flowers: May–September

Plate 378. Giant White-topped Sedge (*Dichromena latifolia*)
Flowers: May–September

Plate 379. Awned Meadow-beauty (*Rhexia aristosa*)
Flowers: June–September

Plate 380. Pipewort (*Eriocaulon decangulare*) Flowers: June–October

Plate 381. Tickseed (*Coreopsis gladiata*) Flowers: August–October

Plate 382. Pond Cypress–Swamp-gum Swamp Forest

Plate 383. Pond Berry (*Lindera melissifolia*)
Flowers: March–April

Plate 384. Pond Berry (*Lindera melissifolia*)
Fruits: August–September

Plate 385. Climbing Fetterbush (*Pieris phillyreifolia*)
Flowers: April–May

Plate 386. Dahoon Holly (*Ilex cassine*) Fruits: October–November

Plate 387. Myrtle Dahoon (*Ilex myrtifolia*) Fruits: October–November

The Ruderal Gardens

Plate 388. Ruderal Garden

Plate 389. Henbit (*Lamium amplexicaule*) Flowers: Mid-Winter–May

Plate 390. Chickweed (*Stellaria media*) Flowers: January–May

Plate 391. Common Blue Violet (*Viola papilionacea*) Flowers: February–May

Plate 392. Common Dandelion (*Taraxacum officinale*)
Flowers: Late February–June

Plate 393. White Clover (*Trifolium repens*)
Flowers: February–September

Plate 394. Mock Strawberry (*Duchesnea indica*)
Flowers: February–Frost

Plate 395. Dewberry (*Rubus trivialis*)
Flowers: March–April

Plate 396. Golden Corydalis (*Corydalis micrantha*)
Flowers: March–April

Plate 397. Cherokee Rose (*Rosa laevigata*)
Flowers: Late March–April

Plate 398. False-garlic (*Allium bivalve*) Flowers: March–May; September–October

Plate 399. Sourgrass (*Rumex hastatulus*) Flowers: March–May

Plate 400. Toadflax (*Linaria canadensis*) Flowers: March–May

Plate 401. Wild-pansy (*Viola rafinesquii*)
Flowers: March–May

Plate 402. Carolina Cranesbill (*Geranium carolinianum*)
Flowers: March–June

Plate 403. Yellow Thistle (*Carduus spinosissimus*)
Flowers: Late March–Early June

Plate 404. Dwarf Dandelion (*Krigia virginica*)
Flowers: Late March–Early June

Plate 405. Prickly Sow-thistle (*Sonchus asper*)
Flowers: March–July

Plate 406. Cut-leaved Evening-primrose (*Oenothera laciniata*)
Flowers: March–July

Plate 407. Moss Verbena (*Verbena tenuisecta*)
Flowers: March–Frost

Plate 408. Lyre-leaved Sage (*Salvia lyrata*)
Flowers: April–May

Plate 409. Prickly Poppy (*Argemone albiflora*)
Flowers: April–May

Plate 410. China-berry (*Melia azedarach*)
Flowers: April–May

Plate 411. Black Locust (*Robinia pseudo-acacia*)
Flowers: April–June

Plate 412. Moth Mullein (*Verbascum virgatum*)
Flowers: April–June

Plate 413. Bigflower Vetch (*Vicia grandiflora*)
Flowers: April–June

Plate 414. Venus' Looking-glass (*Specularia perfoliata*) Flowers: April–June

Plate 415. False-dandelion (*Pyrrhopappus carolinianus*) Flowers: April–June

Plate 416. Crimson Clover (*Trifolium incarnatum*)
Flowers: April–June

Plate 417. Japanese Honeysuckle (*Lonicera japonica*)
Flowers: April–June

Plate 418. Ohio Spiderwort (*Tradescantia ohiensis*)
Flowers: April–July

Plate 419. Chinese Wisteria (*Wisteria sinensis*)
Flowers: April–July

Plate 420. Salt Cedar (*Tamarix gallica*)
Flowers: April–July

Plate 421. Annual Phlox (*Phlox drummondii*)
Flowers: April–July

Plate 422. Black Medic (*Medicago lupulina*)
Flowers: April–August

Plate 423. Heliotrope (*Heliotropium amplexicaule*)
Flowers: April–September

Plate 424. Sour Clover (*Melilotus indica*)
Flowers: April–October

Plate 425. Heal-all (*Prunella vulgaris*)
Flowers: April–Frost

Plate 426. English Plantain (*Plantago lanceolata*)
Flowers: April–Frost

Plate 427. Yarrow (*Achillea millefolium*)
Flowers: April–Frost

Plate 428. Shepherd's Needle (*Bidens pilosa*)
Flowers: April–Frost

Plate 429. Popcorn Tree (*Sapium sebiferum*)
Flowers: May–June

Plate 430. Calliopsis (*Coreopsis basalis*)
Flowers: May–July

Plate 431. Man-of-the-earth (*Ipomoea pandurata*)
Flowers: May–July

Plate 432. Black-eyed Susan (*Rudbeckia hirta*)
Flowers: May–July

Plate 433. Maypops (*Passiflora incarnata*)
Flowers: May–July

Plate 434. Bull Nettle (*Solanum carolinense*)
Flowers: May–July

Plate 435. Mimosa (*Albizia julibrissin*)
Flowers: May–August

Plate 436. Smooth Vetch (*Vicia dasycarpa*) Flowers: May–September

Plate 437. Queen Anne's Lace (*Daucus carota*) Flowers: May–September

Plate 438. Spiny Nightshade (*Solanum sisymbriifolium*) Flowers: May–September

Plate 439. South American Vervain (*Verbena bonariensis*)
Flowers: May–October

Plate 440. Common Purslane (*Portulaca oleracea*)
Flowers: May–October

Plate 441. Pokeweed (*Phytolacca americana*)
Flowers: May–Frost

Plate 442. Bitterweed (*Helenium amarum*)
Flowers: May–Frost

Plate 443. Wood Sage (*Teucrium canadense*)
Flowers: June–August

Plate 444. Woolly Mullein (*Verbascum thapsus*)
Flowers: June–September

Plate 445. Rattle-bush (*Daubentonia punicea*)
Flowers: June–September

Plate 446. Partridge Pea (*Cassia fasciculata*)
Flowers: June–September

Plate 447. Sida (*Sida rhombifolia*)
Flowers: June–October

Plate 448. Common Evening-primrose (*Oenothera biennis*)
Flowers: June–October

Plate 449. Common Nightshade (*Solanum americanum*)
Flowers: June–Frost

Plate 450. Mexican-clover (*Richardia scabra*)
Flowers: June–Frost

Plate 451. Winged Sumac (*Rhus copallina*)
Flowers: July–September

Plate 452. Jimson-weed (*Datura stramonium*)
Flowers: July–September

Plate 453. Rattlebox (*Crotalaria spectabilis*)
Flowers: July–September

Plate 454. Common Morning-glory (*Ipomoea purpurea*)
Flowers: July–September

Plate 455. Common Reed (*Phragmites australis*)
Flowers: July–October

Plate 456. Mexican-tea (*Chenopodium ambrosioides*)
Flowers: July–Frost

Plate 457. Ivy-leaf Morning-glory (*Ipomoea hederacea*) Flowers: July–Frost

Plate 458. Mistflower (*Eupatorium coelestinum*) Flowers: Late July–November

Plate 459. Jacquemontia (*Jacquemontia tamnifolia*) Flowers: August–September

Plate 460. Coastal Morning-glory (*Ipomoea trichocarpa*) Flowers: August–October

Plate 461. Rabbit Tobacco (*Gnaphalium obtusifolium*) Flowers: August–October

Plate 462. Boneset (*Eupatorium perfoliatum*)
Flowers: August–October

Plate 463. Chocolate-weed (*Melochia corchorifolia*)
Flowers: August–October

Plate 464. Gerardia (*Agalinis purpurea*)
Flowers: August–Frost

Plate 465. Red Morning-glory (*Ipomoea coccinea*)
Flowers: August–Frost

Plate 466. Cypress-vine (*Ipomoea quamoclit*)
Flowers: September–Frost

Range-Habitat: Common throughout the state; in the coastal plain in sandy, dry, open woods.

257 **Horse Mint**

Monarda punctata L.
Lamiaceae (Mint Family)
Flowers: July–September
Description: Aromatic, perennial herb with stems 16–40" tall; stems four-angled; leaves opposite; flowers in tight clusters subtended by several wholly or partially pink to lavender bracts; corolla yellow, spotted with purple.
Range-Habitat: Primarily a coastal plain species; common in sandy, dry, open woods and fields.
Comments: Francis Porcher (1869) gives numerous home remedies and medicinal uses of horse mint. There appears to be no medicinal folk use of this plant today in the Lowcountry.

258 **Silver-leaved Grass**

Heterotheca graminifolia (Michaux) Shinners
Synonym: *Chrysopsis graminifolia* (Michaux) Ell.
Asteraceae (Aster or Sunflower Family)
Flowers: July–October
Description: Clumped, silvery-silky, fibrous-rooted, perennial herb; stems erect, 16–32" tall; leaves mostly basal, linear, grasslike, to 14" long, progressively reduced upward.
Range-Habitat: Common throughout the state; in the coastal plain in sandy, dry, open woods; longleaf pine flatwoods and savannas; and pine–mixed hardwood forests.

259 **Prickly-pear; Devil's-tongue**

Opuntia compressa (Salisbury) Macbride
Synonym: *Opuntia humifusa* (Raf.) Raf.
Cactaceae (Cactus Family)
Flowers: August–October
Description: Perennial, woody, fleshy plant with photosynthetic stems; stem segments (joints) tightly held together; large spines present or absent; nodes with tufts of hairlike spines that readily penetrate the skin; leaves alternate, promptly deciduous; fruit a berry, purple to reddish brown when ripe.
Range-Habitat: Common throughout the state; in the coastal plain in sandy, dry, open woods; coastal dunes; and pastures.
Comments: The fruit is edible and can be eaten raw or made into marmalade. It must first be peeled to remove the spines.

260 **Blue Curls**

Trichostema dichotomum L.
Lamiaceae (Mint Family)
Flowers: August–Frost
Description: Bushy-branched annual herb to 32" tall, from a

taproot; main stems spreading-hairy; leaves opposite, entire; branches of the inflorescence paired, dichotomously branched; the four stamens bluish and strongly arched downward.

Range-Habitat: Common throughout the state; in sandy, dry, open woods; sandy fields; and sandy roadsides.

Comments: The common name comes from the distinctive long, curled stamens projecting downward from the flower.

261 **Blazing-star**

Liatris elegans (Walter) Michaux

Asteraceae (Aster or Sunflower Family)

Flowers: September–October

Description: Perennial, erect herb to 5' tall, from a globose, tapered rootstock; stem leaves alternate, linear to narrowly elliptic; bracts below the flower heads dilated, pink.

Range-Habitat: Chiefly coastal plain; common in sandy, dry, open woods and fields.

THE PINELAND GARDENS

262 ***Xeric Sandhills***

263 **Longleaf Pine**

Pinus palustris Miller

Pinaceae (Pine Family)

Female Cones: All Year

Description: Large, evergreen tree; needles 10–16" long; female cones at maturity 6–10" long.

Range-Habitat: Common coastal plain species in sandy soils of longleaf pine flatwoods and savannas, sandhills, pine–mixed hardwood forests, and disturbed sites.

Comments: Longleaf pine, the aristocrat of southern pines, was the dominant pine in the original southern forests of the coastal area. Growing tall and slender to a height of over 120' and living to two hundred to three hundred years, the original longleaf forests quickly fell to the axes of lumberjacks. Natural fires that swept over the cutover areas, feral hogs that consumed the young seedlings, competition from loblolly pine (a more prolific seeder), and the lack of replanting all contributed to longleaf's occupying about 10 percent of its original area today.

Longleaf had numerous uses. It was the basis of the naval stores industry, its trunks were used for masts of sailing ships, and it was used for beams, floors, and general construction. Many of the Lowcountry mansions and plantations are made of "heart pine"—the resinous nature making the wood indestructible by insects and fungi.

264 **Reindeer Moss**
Cladonia evansii Abb.
Description: Shrublike lichen; colonies form discrete, compacted heads.
Range-Habitat: From North Carolina to Texas on sterile, white sand; common in the coastal plain on xeric sandhills.
Comments: Frequently collected for use as imitation shrubbery and trees in model train layouts.

265 **Carolina Ipecac**
Euphorbia ipecacuanhae L.
Euphorbiaceae (Spurge Family)
Flowers: March–May
Description: Perennial, milky herb; from a straight root many feet long, branched near the top into numerous stems with only the tips above ground; above-ground stems smooth, dichotomously branched, forming low, dark green tufts or small mats; a highly variable species.
Range-Habitat: Primarily an inner coastal plain species; in the outer coastal plain reported in Georgetown and Horry counties where it is common in xeric sandhills and sandy, longleaf pine flatwoods.
Comments: Francis Porcher (1869) reported that this plant was "tolerably certain emetic; but liable sometimes to produce excessive nausea by accumulation."

266 **Dwarf Huckleberry**
Gaylussacia dumosa (Andrz.) T. & G.
Ericaceae (Heath Family)
Flowers: March–June
Description: Shrub 4–16" tall with deep, water-holding taproot; leaves alternate with thick cuticle on upper leaf surface, deciduous or semi-evergreen; distinctive, small resinous glands on lower leaf surface; ripe berry black.
Range-Habitat: Occurs in the coastal plain, adjacent piedmont, and mountains; common in the coastal plain in xeric sandhills and sandy, dry, open woods.

267 **Tread-softly; Spurge Nettle**
Cnidoscolus stimulosus (Michaux) Engelm. & Gray
Euphorbiaceae (Spurge Family)
Flowers: Late March–August
Description: Erect or reclining, perennial herb, 6–36" tall; entire plant covered with stinging hairs; leaves alternate, palmately lobed, or dissected.
Range-Habitat: Common throughout the coastal plain and piedmont; in the coastal plain in xeric sandhills; sandy, dry, open woods; sandy, fallow fields; and stable coastal dunes.
Comments: The root is reported to be used as an aphrodisiac in

rural South Carolina; locally it is called the "courage" plant. The stinging hairs can inflict a painful rash on contact and cause a severe reaction in some people.

268 **Gopherweed**
Baptisia perfoliata (L.) R. Brown
Fabaceae (Pea or Bean Family)
Flowers: April–May
Description: Rhizomatous, perennial herb, 20–36" tall; leaves simple and perfoliate; legume matures May–July.
Range-Habitat: Restricted to the coastal plain, primarily the western counties; occasional in xeric sandhills and sandy, dry, open woods.
Comments: Gopherweed can be found in the Tillman Sand Ridge Heritage Preserve and environs in Jasper County.

269 **Gopher Apple**
Chrysobalanus oblongifolius Michaux
Rosaceae (Rose Family)
Flowers: May–June
Description: Low shrub to 16" tall; fire adapted, with an extensive underground stem system; leaves simple, evergreen, alternate, and oblanceolate; fruit matures in September–October.
Range-Habitat: A rare coastal plain species in xeric sandhills.
Comments: Gopher apple has been reported only in the Tillman Sand Ridge Heritage Preserve and environs in Jasper County.

270 **Sandhills Baptisia**
Baptisia cinerea (Raf.) Fernald & Schubert
Fabaceae (Pea or Bean Family)
Flowers: May–June
Description: Rhizomatous, perennial herb, 1–2' tall, with appressed hairs; leaves trifoliolate; legume matures June–August.
Range-Habitat: Chiefly piedmont and coastal plain; in the outer coastal plain, found in Georgetown, Horry, Marion, and Williamsburg counties; occasional in xeric sandhills, longleaf pine flatwoods, and sandy, dry, open woods.

271 **Sandhills Milkweed**
Asclepias humistrata Walter
Asclepiadaceae (Milkweed Family)
Flowers: May–June
Description: Stems stiff and spreading-ascending, one to several from a deep, narrow, tapering root; up to 3' tall; juice milky; leaves opposite, broad, sessile, clasping, usually five to eight pairs; fruits erect, maturing June–July.
Range-Habitat: Primarily a coastal plain species, rare in the

piedmont; in the coastal plain common in xeric sandhills and sandy, dry, open woods.

272 **Beard-tongue**
Penstemon australis Small
Scrophulariaceae (Figwort Family)
Flowers: May–July
Description: Opposite-leaved, perennial herb; stems one to several, 8–28" tall; basal leaves shaped differently from the stem leaves; inflorescence hairy, the hairs simple or glandular.
Range-Habitat: Common throughout the coastal plain and piedmont; in the coastal plain in xeric sandhills; sandy, dry, open woods; dry, fallow fields; and dry, longleaf pine flatwoods.
Comments: The common name comes from the bearded nature of the sterile stamens in some species.

273 **Thistle**
Carduus repandus (Michaux) Persoon
Synonym: *Cirsium repandum* Michaux
Asteraceae (Aster or Sunflower Family)
Flowers: May–July
Description: Deep-rooted, perennial herb, 8–24" tall; stems cobwebby-hairy, leafy to the summit; leaves alternate, covered by numerous, small spines in addition to the scattered, larger ones; heads solitary, or more often terminating short branches from near the summit; bracts below the flower heads are spined tipped.
Range-Habitat: Common throughout the coastal plain; in xeric sandhills; sandy, dry, open woods; and sandy fields.

274 **Warea**
Warea cuneifolia (Muhl.) Nuttall
Brassicaceae (Mustard Family)
Flowers: July–September
Description: Annual herb with erect stems to 4' tall, branched above; leaves alternate, oblanceolate; fruit sickle-shaped, long-stalked, maturing in August–September.
Range-Habitat: Primarily in the xeric sandhills of the midstate; infrequent.
Comments: Warea can be seen at the Tillman Sand Ridge Heritage Preserve and environs in Jasper County.

275 **Soft-haired Coneflower**
Rudbeckia mollis Ell.
Asteraceae (Aster or Sunflower Family)
Flowers: Late August–October
Description: Annual, biennial, or occasionally short-lived perennial; 1–3' tall, with deep taproot; stems freely branched with dense, spreading hairs; stem leaves alternate, sessile, with dense hairs.

Range-Habitat: A coastal plain species; found only in Jasper County in the Tillman Sand Ridge Heritage Preserve and environs; xeric sandhills and sandy, dry, open woods; rare.

276 **Summer-farewell**
Petalostemum pinnatum (Walter ex J. F. Gmelin) Blake
Fabaceae (Pea or Bean Family)
Flowers: August–Frost
Description: Perennial, aromatic herb, 12–24" tall from a taproot; stems smooth, branched above; leaves odd-pinnately compound, glandular-dotted, with three to eleven leaflets.
Range-Habitat: A coastal plain species; mostly in the southwest coastal plain, but also in Marion County; occasional in xeric sandhills and sandy, dry, open woods.
Comments: Summer-farewell is very unlike most members of the pea family in appearance. Good populations exist in the Tillman Sand Ridge Heritage Preserve and environs.

277 **Turkey Oak**
Quercus laevis Walter
Fagaceae (Beech Family)
Fruits: September–October
Description: Small tree; bark thick with deep, irregular furrows and scaly, rough ridges; inner bark reddish; leaves alternate, deciduous, generally with three wide-spreading main lobes (like a turkey's foot) that are pointed and bristle-tipped.
Range-Habitat: Common in the coastal plain; in xeric sandhills; sandy, longleaf pine flatwoods; and other sites with poor, sandy soil.
Comments: This is one of the few oaks that will grow in sandy, sterile soil. The tree is not large enough for lumber, but the seasoned wood makes an excellent firewood. It is often used as the fuel in pits for "southern barbecue" of hogs.

Leaves of young seedlings are orientated at right angles to the ground; this reduces heat absorption from direct sunlight and reflected from bare white sand.

278 **Woolly Golden-aster**
Heterotheca gossypina (Michaux) Shinners
Synonym: *Chrysopsis gossypina* Nuttall
Asteraceae (Aster or Sunflower Family)
Flowers: September–October
Description: Biennial or short-lived perennial, 12–28" tall; reclining, ascending, or erect; densely woolly-hairy throughout, commonly extending to the bracts below the flower heads.
Range-Habitat: Occasional throughout the coastal plain in xeric sandhills; sandy, dry, open woods; and sometimes on beaches.

279 **Gerardia**

Agalinis setacea (J. F. Gmelin) Raf.
Synonym: *Gerardia setacea* (Walter) J. F. Gmelin
Scrophulariaceae (Figwort Family)
Flowers: September–October
Description: Dark green or reddish annual, 8–24" tall; leaves long and very slender; corolla throat with yellow lines.
Range-Habitat: Chiefly coastal plain but infrequent in lower South Carolina; occurs in xeric sandhills, sandy pinelands, and sandy, dry, open woods.
Comments: All species of *Agalinis* are thought to be parasitic (to various degrees) on roots of grasses or other herbs.

Seven other species of *Agalinis* occur in the coastal plain. The reader may refer to the standard floras for descriptions of the various species. *A. setacea* is the only species to grow in the extreme xeric sandhills.

280 **Rose Dicerandra**

Dicerandra odoratissima Harper
Lamiaceae (Mint Family)
Flowers: September–October
Description: Freely branched, aromatic herb with appressed hairs; erect, with four-angled stems, 8–20" tall; leaves opposite, linear to narrowly elliptic; flowers pink to lavender in our area.
Range-Habitat: Rare coastal plain species; xeric sandhills and sandy, dry, open woods.
Comments: Rose dicerandra is known only in the Tillman Sand Ridge Heritage Preserve and environs in Jasper County.

281 **Rosemary**

Ceratiola ericoides Michaux
Empetraceae (Crowberry Family)
Flowers: October–November
Description: Much-branched, dense shrub, 2–5' tall, with four rows of short, slender, needlelike leaves; male and female flowers on separate plants, sessile in leaf axils.
Range-Habitat: Rare and scattered in the coastal plain; limited to xeric sandhills.
Comments: This is not the herb rosemary. Small (1933) states that throughout its range, fire has destroyed vast areas of *Ceratiola*.

282 *Longleaf Pine Flatwoods*

283 **Cinnamon Fern**

Osmunda cinnamomea L.
Osmundaceae (Royal Fern Family)
Fruits: March–May
Description: Perennial from stout, woody rhizomes, without scales; rhizomes buried in massive clump of old leaf stalk

bases and fibrous, black roots; leaves erect, around 3' tall; fertile leaves appear in the spring bearing the cinnamon-colored sporangia, then wither; sterile leaves arise from the rootstocks in a circle to the outside of the fertile leaves.

Range-Habitat: Cinnamon fern is common in the mountains and coastal plain, sporadic in the piedmont; in the outer coastal plain it occurs in longleaf pine flatwoods and savannas, cypress swamps, ditches, and freshwater marshes.

Comments: Cinnamon fern transplants well into home gardens as long as the soil is moist and the site shady. In the spring the young fiddleheads (young leaves) may be cooked as a vegetable. Its matted rootstocks are used as a culture medium for orchids. The common name comes from the rusty-wool hairs on the young leaves.

284 **Red Chokeberry**

Sorbus arbutifolia (L.) Heynold var. *arbutifolia*
Synonym: *Aronia arbutifolia* (L.) Ell.
Rosaceae (Rose Family)
Flowers: March–May
Description: Medium-sized shrub, often forming large colonies; leaves with fine, gland-tipped teeth; leaves persistently gray-felted beneath, dark green above; fruit red, astringent, maturing September–November, long persistent.
Range-Habitat: Common throughout the state; in the outer coastal plain common in longleaf pine flatwoods and savannas and pocosin borders.

285 **Dwarf Azalea**

Rhododendron atlanticum (Ashe) Rehder
Ericaceae (Heath Family)
Flowers: April–May
Description: Deciduous, low, erect shrub, 3–5' tall; often forming large colonies from rhizomes; flowers appearing before the leaves; corolla white, lavender to deep pink; corolla sticky from glands.
Range-Habitat: Primarily a coastal plain species; occasional in pine flatwoods and pine savannas, less so in xeric sandhills.

286 **Stagger-bush**

Lyonia mariana (L.) D. Don
Ericaceae (Heath Family)
Flowers: April–May
Description: Rhizomatous shrub to 5' tall; leaves deciduous to semi-evergreen.
Range-Habitat: A coastal plain and adjacent piedmont species; common in the coastal plain in longleaf pine flatwoods and sandy, dry, open woods.

287 **White-topped Aster**

Aster reticulatus Pursh
Asteraceae (Aster or Sunflower Family)

Flowers: Late April–Early June
Description: Perennial with few to many stems forming clumps from a branching, thickened base; 16–36" tall.
Range-Habitat: Reported only in Jasper and Beaufort counties in South Carolina; grows in longleaf pine and pine–saw palmetto flatwoods; common in Georgia and Florida, but no reference could be found on its status in South Carolina.
Comments: This is a very conspicuous and robust aster which is easily seen from a distance in the pine flatwoods, often forming dense colonies.

288 **Male-berry; He-huckleberry**
Lyonia ligustrina (L.) DC.
Ericaceae (Heath Family)
Flowers: April–June
Description: Rhizomatous, much-branched, deciduous shrub; 3–12' tall; corolla globular, white, or pale rose; leaves alternate, minutely toothed.
Range-Habitat: Common throughout the state; in the coastal plain it grows in moist areas in longleaf pine flatwoods and oak-hickory and pine–mixed hardwood forests.

289 **Sundrops**
Oenothera fruticosa L.
Onagraceae (Evening-primrose Family)
Flowers: April–August
Description: Perennial, hairy herb; stems erect, to 32" tall; capsule club-shaped.
Range-Habitat: Common throughout the piedmont and coastal plain; common in the outer coastal plain in longleaf pine flatwoods and savannas and roadsides.
Comments: Sundrops is representative of the group of evening-primroses that bloom during the day; members of the other group bloom mostly in the evening. This species is commonly cultivated as an ornamental in gardens.

290 **Yellow False-indigo; Wild Indigo; Horse-fly Weed**
Baptisia tinctoria (L.) R. Brown
Fabaceae (Bean or Pea Family)
Flowers: April–August
Description: Bushy-branched, rhizomatous perennial up to 3' tall; usually blackening upon drying; leaves alternate, palmately compound with three leaflets.
Range-Habitat: Throughout the state; common in the outer coastal plain in longleaf pine flatwoods and open, thin woods.
Comments: Wild indigo was used and cultivated in colonial times as a source of dye; it was not a quality dye and could not replace true indigo obtained from Asian and tropical members of the genus *Indigofera*. The specific epithet *tinctoria* refers to tincture, meaning "a dyeing substance." Francis Porcher (1869) gives a long discourse on the use of

this plant as a medicine. He also relates how the fresh plant was attached to the harness of horses to keep flies off—hence, one of its common names. He further states: "I have noticed that they [flies] will not remain upon the plant."

291 **American Chaff-seed**
Schwalbea americana L.
Synonym: *Schwalbea australis* Pennell
Scrophulariaceae (Figwort Family)
Flowers: May–June
Description: Erect, hairy, unbranched, perennial herb, 1–2' tall; corolla yellow or purplish; leaves alternate, the largest at the base and gradually diminishing in size upwards.
Range-Habitat: A rare coastal plain species; it grows in open, grassy-like areas in longleaf pine flatwoods maintained by frequent fires.
Comments: Chaff-seed is a Federally Endangered Species. It is rare throughout the Southeast. Forty-two extant populations occur in South Carolina. Some of the historic populations have been lost due to habitat destruction. Chaff-seed is a root-parasite with the host species representing components of the habitat.

292 **Black-root**
Pterocaulon pycnostachyum (Michaux) Ell.
Asteraceae (Aster or Sunflower Family)
Flowers: May–June
Description: Perennial herb from a black, tuberous, thickened root; stems erect, 16–32" tall with dense, matted hairs; stem winged by the base of the leaf stalks extending down the stem; vegetative parts often remain through the winter.
Range-Habitat: A coastal plain species; common in longleaf pine flatwoods and sandy fields.
Comments: The root was used in a variety of folk remedies according to Francis Porcher (1869) and Morton (1974). The latter states that even today a decoction of the root is taken for colds and menstrual cramps, or the whole root is boiled and the tea taken for backache.

293 **Goat's Rue; Devil's Shoestrings; Hoary-pea**
Tephrosia virginiana (L.) Persoon
Fabaceae (Bean or Pea Family)
Flowers: May–June
Description: Perennial herb; 8–28" tall; densely grayish hairy—hence, the common name hoary-pea; leaves odd-pinnately compound, covered with grayish hairs.
Range-Habitat: Goat's rue is common throughout the state; in the outer coastal plain it is found in longleaf pine flatwoods, roadsides, and in sandy, dry, open woods.
Comments: Francis Porcher (1869) states that goat's rue was used by the Indians (but did not state how). He further adds

it was later used as a vermifuge (but not by whom). The reader may refer to his book for additional material on its historic use.

The common name Devil's shoestrings comes from its long, stringy roots. It was once fed to goats to increase their milk production; when it was discovered the roots contain rotenone, the practice was discontinued. Rotenone is used as an insecticide and fish poison, but it is not poisonous to mammals.

294 **Phlox**
Phlox carolina L.
Polemoniaceae (Phlox Family)
Flowers: May–July
Description: Perennial herb; stems erect, solitary, or clustered; stems to 3' tall.
Range-Habitat: Common throughout the state; common in the outer coastal plain in longleaf pine flatwoods, oak-hickory forests, clearings, and roadsides.

295 **White Wild Indigo**
Baptisia alba (L.) R. Brown
Fabaceae (Bean or Pea Family)
Flowers: May–July
Description: Rhizomatous, perennial herb to 4' tall; leaves trifoliolate; flowers in several, terminal racemes.
Range-Habitat: Chiefly a piedmont species but still widespread in the coastal plain; common in longleaf pine flatwoods.

296 **Curly Milkweed; Clasping Milkweed**
Asclepias amplexicaulis Smith
Asclepiadaceae (Milkweed Family)
Flowers: May–July
Description: Perennial herb, one to three stems from a root crown; 16"–3' tall; juice milky; leaves opposite, margins wavy-crisped, auriculate-clasping.
Range-Habitat: Common throughout the state; in the outer coastal plain, common in longleaf pine flatwoods and sandy, dry, open woods.

297 **Sweet Pepperbush**
Clethra alnifolia L.
Clethraceae (White Alder Family)
Flowers: May–July
Description: Shrub 3–10' tall; flowers fragrant; spreads by underground stems and forms clumps; leaves wedged-shaped; capsules ripen September–October, remaining on plant long after flowering which helps to identify it in the winter.
Range-Habitat: A coastal plain plant; common in longleaf pine flatwoods, pocosins, pond cypress savannas and pond

cypress swamp forests.

Taxonomy: Separated into two varieties by some authors: *C. alnifolia* var. *alnifolia* with hairless leaves and *C. alnifolia* var. *tomentosa* with leaves densely hairy beneath. Both varieties occur in the coastal plain with overlapping ranges and in the same habitats.

Comments: In pond cypress savannas and pond cypress swamps, the plant grows on cypress knees and on buttresses.

298 **Lead Plant**

Amorpha herbacea Walter
Fabaceae (Bean or Pea Family)
Flowers: May–July
Description: Herb 1–5' tall; leaves odd-pinnately compound.
Range-Habitat: Primarily a coastal plain species; common in longleaf pine flatwoods and open fields.

299 **Sneezeweed**

Helenium flexuosum Raf.
Asteraceae (Aster or Sunflower Family)
Flowers: May–August
Description: Perennial herb, 16–36" tall; leaves alternate, with the base of the blades extending down the stems making them winged.
Range-Habitat: Essentially throughout the state; common in the outer coastal plain in longleaf pine flatwoods, ditches, and alluvial pastures.
Comments: The common name comes from the use of its dried leaves in making snuff which is inhaled to cause sneezing in the belief it would rid the body of evil spirits.

300 **Rattlesnake Master; Button Snakeroot**

Eryngium yuccifolium Michaux
Apiaceae (Parsley Family)
Flowers: June–August
Description: Perennial herb 2–6' tall; stem solitary, erect, branched above; flowers in heads, developing a bluish cast at maturity; the long, narrow leaves suggest a monocot, but it is a dicot.
Range-Habitat: Common throughout coastal plain, piedmont, and lower mountains; common in the outer coastal plain in longleaf pine flatwoods, sandy roadsides, and sandy, dry, open woods.
Comments: The common names refer to the supposed ability to cure snakebite, or perhaps to keep rattlesnakes away. Native Americans used the plant for a variety of folk remedies as did the early settlers.

301 **Pencil Flower**

Stylosanthes biflora (L.) BSP.
Fabaceae (Bean or Pea Family)

Flowers: June–August

Description: Perennial herb with one to several prostrate to erect stems, 5–20" long; leaves trifoliolate; legume consists of two portions, the lower frequently aborting and becoming stalklike; legumes maturing July–October, the seeds eaten by birds.

Range-Habitat: Common throughout the state; in the outer coastal plain it grows in longleaf pine flatwoods, roadsides, waste places, and sandy, dry, open woods.

302 **Ironweed**

Vernonia acaulis (Walter) Gleason

Asteraceae (Aster or Sunflower Family)

Flowers: Late June–August

Description: Perennial herb; stems 2–4' tall, arising from a crown; principal leaves at the base of the plant, those higher on the stem, if any, much smaller.

Range-Habitat: A coastal plain and piedmont species; common in longleaf pine flatwoods and sandy, dry, open woods.

303 **Whorled-leaf Coreopsis; Wood Tickseed**

Coreopsis major Walter

Asteraceae (Aster or Sunflower Family)

Flowers: June–August

Description: Perennial herb, 20–40" tall; rhizomatous, but with stems commonly tufted; upper and middle nodes with opposite leaves, deeply palmately divided into three segments, giving the appearance of six leaves in a whorl.

Range-Habitat: Common throughout the state; most common in longleaf pine flatwoods, less so in oak-hickory forests and sandy, dry, open woods.

Taxonomy: Different varieties have been described by various authors. The reader may refer to the manuals listed in the bibliography for additional information.

304 **Blazing-star**

Liatris squarrosa (L.) Michaux

Asteraceae (Aster or Sunflower Family)

Flowers: June–August

Description: Perennial herb from a thickened rootstock; 12–32" high; leaves linear, those near the base 3–10" long; tips of the bracts below the flowering heads pointed away from the heads.

Range-Habitat: Throughout the state; occasional in the coastal plain in sandy, longleaf pine flatwoods and sandy, oak-hickory forests.

305 **Bicolor**

Lespedeza bicolor Turcz.

Fabaceae (Bean or Pea Family)

Flowers: June–September

Description: Woody, perennial shrub, 3–10' tall; often dies back to the ground in winter; leaves trifoliolate, leaflets green above, paler below; legumes maturing August–November.

Range-Habitat: Common in the piedmont and coastal plain; in the coastal plain in longleaf pine flatwoods, fields, roadsides, and fence rows.

Comments: Bicolor was introduced from Japan and planted on quail plantations because its seeds are available during February–March when native seeds are at a minimum. The plant has escaped from cultivated sites and appears to spread more readily in sites with high fire frequency.

The common name refers to the two-colored leaves.

306 **Spiked Medusa**

Eulophia ecristata (Fernald) Ames

Orchidaceae (Orchid Family)

Flowers: June–September

Description: Perennial herb from a thickened corm; leaves several, arising from the corm, linear-lanceolate, 6–28" long, two of which dominate; single flowering stem grows from the corm near the leaves, 20"–5.5' tall; by November the capsules have matured and spread their seeds.

Range-Habitat: A rare coastal plain species; reported in Berkeley, Hampton, Dorchester, and Charleston counties; in grassy areas in longleaf pine flatwoods and savannas.

Comments: Little is known about this species in South Carolina. It may be present in a site one year, then disappear for several years (as do many orchids). It is easily overlooked because it grows among tall grasses such as *Andropogon* ssp. and *Paspalum* ssp.

307 **Dollar-weed**

Rhynchosia reniformis DC.

Fabaceae (Bean or Pea Family)

Flowers: June–September

Description: Erect, perennial herb, 2–10" tall; leaves kidney-shaped, covered with resinous dots.

Range-Habitat: Common coastal plain species; in the outer coastal plain in longleaf pine flatwoods and sandy, dry, open woods.

308 **Sensitive Brier**

Schrankia microphylla (Solander ex Smith) Macbride

Fabaceae (Bean or Pea Family)

Flowers: June–September

Description: Perennial herb with prostrate to weakly arching stems, often 3–6' long; leaves, stems, and fruits covered with prickles; leaves deciduous, even-bipinnately compound; flowers in heads.

Range-Habitat: Common throughout the state; in the outer coastal plain in longleaf pine flatwoods, roadsides, and

sandy, dry, open woods.
Comments: The common name refers to leaves that fold quickly when touched.

309 **St. Peter's-wort**
Hypericum stans (Michaux) P. Adams & Robson
Hypericaceae (St. John's-wort Family)
Flowers: June–October
Description: Erect shrub, 1–3' tall; stems with shredded, old bark; sepals four and unequal, the two outer, larger ones enclosing the two much narrower, inner ones.
Range-Habitat: Chiefly coastal plain and lower piedmont; common in outer coastal plain in longleaf pine flatwoods and savannas, ditches, and sandy, dry, open woods.

310 **Golden-aster**
Heterotheca mariana (L.) Shinners
Synonym: *Chrysopis mariana* (L.) Ell.
Asteraceae (Aster or Sunflower Family)
Flowers: Late June–October
Description: Fibrous-rooted perennial from a short, woody rhizome; stems erect, 8–32" tall; loosely woolly-hairy when young; leaves mostly basal, to 5" long; stem leaves reduced upward.
Range-Habitat: Common throughout the state; in the outer coastal plain in longleaf pine flatwoods and old fields.

311 **Galactia**
Galactia macreei M. A. Curtis
Fabaceae (Pea or Bean Family)
Flowers: July–September
Description: Herbaceous perennial, climbing or twining vine; petals pink to purple; leaves trifoliolate; legume appressed-hairy, dehiscent with laterally twisting valves.
Range-Habitat: Primarily an outer coastal plain species; common in longleaf pine flatwoods and savannas and pine–mixed hardwood forests.
Taxonomy: Duncan and Duncan (1987) include *G. macreei* as a synonym of *Galactia volubilis* (L.) Britton; Radford et al. (1968) list the two as separate species, with the former primarily a coastal species and the latter occurring statewide. The difference between the two, if indeed they are separate species, is based on the type of hairs. The reader may refer to the above references for a more detailed account of their taxonomy.

312 **Broad-leaved Eupatorium**
Eupatorium rotundifolium L.
Asteraceae (Aster or Sunflower Family)
Flowers: August–October

Description: Perennial herb, 2–5' tall; stems mostly single or paired from a short rhizome; plant hairy throughout; leaves opposite, sessile, or nearly so.

Range-Habitat: Throughout the state; common in the coastal plain, less frequent in the piedmont, rare in the mountains; in the coastal plain it occurs in longleaf pine flatwoods and savannas.

Taxonomy: Numerous varieties are recognized. The reader may refer to Cronquist (1980) for a review.

313 **Deer's-tongue**

Trilisa paniculata (Walter ex J. F. Gmelin) Cassini
Synonym: *Carphephorus paniculatus* (J. F. Gmelin) Herbert
Asteraceae (Aster or Sunflower Family)
Flowers: August–October

Description: Perennial herb 16–32" tall; basal leaves narrowly elliptic to oblanceolate, 3–9" long; stem leaves much reduced, numerous, erect, and clothing the stem.

Range-Habitat: A coastal plain species; common in longleaf pine flatwoods and savannas.

314 **Thistle**

Carduus virginianus L.
Synonym: *Cirsium virginianum* (L.) Michaux
Asteraceae (Aster or Sunflower Family)
Flowers: August–Frost

Description: Slender biennial, 20–40" tall, roots tuberous-thickened; leaf divisions spine-tipped; leaves white-hairy beneath; bracts below the flowering heads spine-tipped.

Range-Habitat: A coastal plain plant; common in longleaf pine savannas and flatwoods.

Comments: Some argue against including the thistles as wildflowers because of their weedy nature in fields, pastures, and gardens; but to view the native species in their natural gardens is to gain a different appreciation of their beauty.

315 **Stiff-leaved Aster**

Aster linariifolius L.
Asteraceae (Aster or Sunflower Family)
Flowers: Late September–November

Description: Perennial herb; stems one to several from the base, wiry, 8–18" tall; leaves numerous, linear, rigid, the lowest ones soon deciduous.

Range-Habitat: Throughout the state but more common in the coastal plain; common in outer coastal plain in longleaf pine flatwoods and sandy, dry, open woods.

316–317 ***Longleaf Pine Savannas***

318 **Pixie-moss**

Pyxidanthera barbulata Michaux
Diapensiaceae (Pixie Family)

Flowers: March–April

Description: Creeping, perennial subshrub with evergreen, lanceolate leaves about .25" long; flowers sessile, about .25" across.

Range-Habitat: Rare and irregularly distributed in the northern coastal plain; longleaf pine savannas, xeric woodlands, and pocosin ecotones.

Comments: Good populations of pixie-moss in South Carolina are found in Cartwheel Bay Heritage Preserve and the Vaughn Tract of the Little Pee Dee River Heritage Preserve in Horry County.

319 **Sun-bonnets**

Chaptalia tomentosa Vent.

Asteraceae (Aster or Sunflower Family)

Flowers: March–May

Description: Fibrous-rooted, perennial herb with a basal cluster of leaves; leaves densely white-hairy beneath, glossy green above; flowering stems one to several, 3–16" tall; heads solitary, nodding at first, then erect, and again nodding after flowering.

Range-Habitat: A coastal plain species common in longleaf pine savannas and flatwoods, roadsides, and sandhill ecotones.

320 **Bearded Grass-pink**

Calopogon barbatus (Walter) Ames

Orchidaceae (Orchid Family)

Flowers: March–May

Description: Erect perennial, arising from a corm; leafless, flowering stalk 6–18" tall; leaves (when present) one or two, narrowly linear and grasslike, sheathing the flowering stalk at the base; flowers rose-pink, rarely white, mostly opening simultaneously; like all species of *Calopogon,* the lip is the uppermost segment of the perianth and is dilated and bearded.

Range-Habitat: A coastal plain orchid widely scattered throughout; infrequent in longleaf pine savannas and flatwoods.

Comments: Although the flowering period is given as March–May, its actual period of being in flower for any one given year is probably two to four weeks; at least, this has been my observation. In addition, it is sensitive to changes in environmental parameters; one year it may be scarce, the next year much more common.

321 **Yellow Butterwort**

Pinguicula lutea Walter

Lentibulariaceae (Bladderwort Family)

Flowers: March–May

Description: Perennial, carnivorous plant to 18" tall; upper leaf surface covered with short, glandular hairs in which insects become mired; leaves in a basal cluster and retained over the winter.

Range-Habitat: Common throughout the coastal plain; occurs in longleaf pine flatwoods and savannas.

Comments: The genus name derives from the Latin *pinguis* which means "somewhat fat" and refers to the greasy texture of the leaf. The butterworts were used as crude bandages during the Civil War. The plant was inverted and the upper surface placed on the wound.

322 **Lance-leaved Violet**

Viola lanceolata L.
Violaceae (Violet Family)
Flowers: March–April

Description: Perennial herb, without a leafy stem; leaves arise from the rhizome, are narrowly to widely lanceolate or linear, with a long wedge-shaped base.

Range-Habitat: Throughout the coastal plain; common in longleaf pine savannas and wet sites in longleaf pine flatwoods.

Similar Species: *V. primulifolia* L. also has white flowers but has leaves ovate to elliptic; similar habitat, but throughout the state.

323 **Leopard's-bane**

Arnica acaulis (Walter) BSP.
Asteraceae (Aster or Sunflower Family)
Flowers: Late March–Early June

Description: Perennial herb; stems erect, 6–32" tall, usually one from a crown; principal leaves basal, forming a cluster; stem leaves few and reduced; both type leaves glandular-hairy.

Range-Habitat: Throughout the coastal plain and lower piedmont; in the coastal plain, common in longleaf pine savannas and flatwoods, roadsides, and open woodlands.

324 **Hooded Pitcher-plant**

Sarracenia minor Walter
Sarraceniaceae (Pitcher-plant Family)
Flowers: April–May

Description: Carnivorous, flowering herb; perennial, evergreen, rhizomatous; leaves 6–24" tall, modified into hollow, tubular structures to catch insects (and small mammals in some species); leaves winged; hood arching closely over the opening of the pitcher; upper portion of hood spotted with white or translucent blotches (the windows); flowering stalk usually shorter than the leaves.

Range-Habitat: A coastal plain species; frequent in longleaf pine savannas and wet ditches, less frequent in longleaf pine flatwoods.

Comments: Hooded pitcher-plant gets it common name from the hood arching over the opening into the pitcher. The arching hood acts to keep flying insects from exiting the pitcher. After an insect has entered the pitcher and drunk its fill on the nectar roll, it attempts to exit. The hood blocks most of the light entering the pitcher, but the insect still seeks escape by flying toward the brightest light which comes from the back of the pitcher. Here, pigment-free areas (the windows) allow some light to enter. As the insect tries to fly through the windows, it repeatedly bumps into the back of the hood, becomes stunned, and ultimately falls down the pitcher and drowns in the mixture of water and enzymes at the base.

Morton (1974) reports an unusual folk remedy with hooded pitcher-plant still practiced in the Lowcountry: "Rootstock is boiled and the decoction kept in a jar, applied warm on skin rashes or eruptions. People say the spots on the leaves are a sign that the plant is a good remedy for skin troubles, a belief which harks back to the old 'Doctrine of Signatures.'"

325 **Indian Paint Brush**

Castilleja coccinea (L.) Sprengel
Scrophulariaceae (Figwort Family)
Flowers: April–May
Description: Erect, hairy, hemiparasitic annual or biennial herb; to 28" tall, with one unbranched stem from a basal cluster of leaves; flowers variable, yellow to greenish yellow; the leaflike bracts, subtending the flowers, vary from scarlet to yellow.
Range-Habitat: Indian paint is primarily found in the mountains and piedmont where it is infrequent; however, several disjunct populations have been found in the coastal plain in Berkeley and Williamsburg counties where it is very rare in meadows and longleaf pine savannas.
Comments: The beauty of this plant comes not from the small flowers but from the brightly tipped bracts. The common name comes from an Indian legend that discarded brushes used by a brave to paint a brilliant sunset grew into flowers. Indians used the roots, mixed with iron minerals, to dye deer skins black.

326 **Violet Butterwort**

Pinguicula caerulea Walter
Lentibulariaceae (Bladderwort Family)
Flowers: April–May
Description: Carnivorous plant; similar to *P. lutea* (plate 321) except this species has a blue corolla.
Range-Habitat: A coastal plain species, somewhat more widespread and more common than yellow butterwort; longleaf pine savannas and moist, longleaf pine flatwoods.
Comments: Same as for *P. lutea.*

327 **Sundew**

Drosera leucantha Shinners
Droseraceae (Sundew Family)
Flowers: April–May
Description: Carnivorous herb with leaves in a basal cluster; leaf blades covered with stalked glands that exude a clear, sticky material that aids in catching insects; flowering stalk glandular-hairy, .8–2.5" tall.
Range-Habitat: Throughout the coastal plain; common in longleaf pine savannas, moist longleaf pine flatwoods, sandy roadside ditches, and seepage bogs.
Similar Species: *D. capillaris* Poiret differs in having a flowering stalk without glands and hairs, and the leaf blades more long than wide; habitat and range of both are similar.
Comments: A good habitat to locate sundews is along the bank of earth pushed up by fire-plows in pine savannas.

The common name of the sundews comes from the sticky, dew-covered tentacles on the leaves that shine and glitter in the early morning with the color of the sun's spectrum.

328 **Black Snakeroot; Crow-poison**

Zigadenus densus (Desr.) Fernald
Liliaceae (Lily Family)
Flowers: April–Early June
Description: Slender herb from a bulblike base; stems to 5' tall; leaves mostly basal, elongate-linear, one to three, enclosed by a purplish, bladeless sheath.
Range-Habitat: Throughout the coastal plain; common in longleaf pine savannas and pocosins.
Comments: Although all species of *Zigadenus* are considered poisonous, much variation occurs in their toxicity. Reported poisoning is mainly among grazing livestock; however, cases of human poisoning have been reported. *Z. densus* is considered particularly potent. The reader may refer to Kingsbury (1964) for more detail.

329 **Colicroot**

Aletris farinosa L.
Liliaceae (Lily Family)
Flowers: Late April–Early June
Description: Perennial herb with thick, short rhizomes; leaves mostly basal, rather narrowly lanceolate; stem leaves greatly reduced; flower stalk 1–3' tall; perianth covered with small, sticky projections giving a mealy appearance.
Range-Habitat: Throughout the state; in the coastal plain common in longleaf pine savannas and flatwoods, roadsides, and old fields.
Comments: This plant and *A. aurea* (plate 340) were used medicinally to treat colic in colonial times. Historically, an infusion of the roots in vinegar was used as a purgative. Morton's (1974) research indicates rural people still chew

the rootstock of both species to stop toothache.

The generic name *Aletris* is Greek for "a female slave who grinds corn," in reference to the apparent mealy texture of the flowers.

330 **Yellow Meadow-beauty**
Rhexia lutea Walter
Melastomataceae (Meadow-beauty Family)
Flowers: April–July
Description: Perennial herb from a stout, woody taproot or rootstock; stems to 2' tall, usually branched, glandular-hairy.
Range-Habitat: A coastal plain species; common in longleaf pine savannas and low, longleaf pine flatwoods.
Comments: This is the only species of meadow-beauty with yellow flowers.

331 **Common Grass-pink**
Calopogon tuberosus (L.) BSP.
Synonym: *Calopogon pulchellus* (Salisbury) R. Brown
Orchidaceae (Orchid Family)
Flowers: April–July
Description: Perennial herb, arising from a corm; leafless, flowering stalk 6–24" tall; one or two leaves (when present), narrowly linear and grasslike, sheathing the flowering stalk at the base; flowers open successively up the raceme; each flower lasts for about five days, and often there are three or four open at once; lip strongly dilated and bearded with numerous hairs in the center; flowers pink to rose-purple or magenta-crimson, rarely white.
Range-Habitat: Throughout the coastal plain and mountains; rare in the piedmont; in the coastal plain it occurs in longleaf pine savannas, moist longleaf pine flatwoods, meadows, and seepage areas on pocosin borders.
Similar Species: In the same habitat with bearded grass-pink and common grass-pink is pale grass-pink (*Calopogon pallidus*). It differs from bearded grass-pink in having lateral sepals reflexed, and from common grass-pink in having the middle sepal less than .8" long compared to more than .8" long for common grass-pink. The flowers of pale grass-pink are pale pink with pure white not uncommon.
Comments: A most unusual feature of the grass-pinks is the inverted perianth where the lip is uppermost. (In all other orchid genera, the lip is lowermost.) The upper lip acts like an elevator: the weight of an insect that lands on the lip causes it to bend downward to bring the insect in contact with the reproductive structure below called the column. Cross-pollination is thus effected.

The grass-pinks have no nectar to entice insects. Instead they mimic flowers that do have nectaries. Growing in association with the coastal grass-pinks is smooth meadow-beauty (*Rhexia alifanus,* plate 343), with pink petals and long yellow stamens that provide nectar and food for visiting insects. The pink petals and yellow hairs (the beard) of the lip, mimicking the pink petals and yellow stamens of the meadow-beauty, probably fool insects into mistaking the orchid flower for the meadow-beauty.

332 **Blue-hearts**

Buchnera floridana Gandoger
Scrophulariaceae (Figwort Family)
Flowers: April–October
Description: Perennial herb with erect, simple stems, 16–32" tall; probably root-parasitic; leaves opposite, three-veined, somewhat reduced above; entire plant turns black upon drying; corolla purple to white.
Range-Habitat: A common coastal plain species; found in longleaf pine savannas, roadsides, and meadows.

333 **Orange Milkwort**

Polygala lutea Ell.
Polygalaceae (Milkwort Family)
Flowers: April–October
Description: Biennial or short-lived perennial to 16" tall; stems usually unbranched, smooth, one to several from base; peak blooming occurs in spring.
Range-Habitat: Common throughout the coastal plain; grows in longleaf pine savannas and flatwoods, roadsides, and pocosins.
Comments: The specific epithet *lutea* is Latin for "yellow" and refers to the flower's color when dried; when fresh the flowers are bright orange.

334–335 **Venus' Fly Trap**

Dionaea muscipula Ellis
Dionaeaceae (Venus' Fly Trap Family)
Flowers: May–June
Description: Perennial, carnivorous herb from a short rhizome; flowering stalk 4–12" tall; leaves in basal clusters, modified into trapping structures consisting of two hinged, sensitive lobes with stout, marginal bristles; each lobe contains trigger hairs by which the trap can be sprung; almost entire upper surface of lobes covered with glands of two kinds: nectar producing and enzyme producing.
Range-Habitat: Endemic to the coastal plain of the Carolinas; in South Carolina it occurs only in Horry and Georgetown counties; infrequent in wet, sandy ditches; longleaf pine savannas; and sphagnum openings in pocosins.
Comments: Venus' fly trap is protected by law in South Carolina. Good populations of Venus' fly trap occur in Lewis Ocean Bay Heritage Preserve and Cartwheel Bay Heritage Preserve in Horry County.

336 **Milkweed**

Asclepias longifolia Michaux
Asclepiadaceae (Milkweed Family)
Flowers: May–June
Description: Perennial herb with milky juice; stems simple, 8–27" tall; leaves opposite to subopposite, linear; corolla greenish white; fruits erect, maturing in June.
Range-Habitat: A coastal plain species; common in longleaf pine savannas.

337 **Spreading Pogonia; Rosebud Orchid**

Cleistes divaricata (L.) Ames
Orchidaceae (Orchid Family)
Flowers: May–July
Description: Perennial herb, 1–2' tall; leaf solitary, inserted above the middle of the stem; sepals widely spreading; entire plant has a bluish green color with a fine, frosty white coating.
Range-Habitat: Chiefly a coastal plain orchid but also in the mountains; in the coastal plain infrequent in longleaf pine savannas and flatwoods, swamps, and openings in pocosins.

338 **Narrow-leaved Skullcap; Hyssop Skullcap**

Scutellaria integrifolia L.
Lamiaceae (Mint Family)
Flowers: May–July
Description: Perennial herb with erect, four-angled stems, 1–2.5' tall; in clumps of one to several stems; calyx with prominent hump on upper side.
Range-Habitat: Common throughout the state; in the coastal plain, common in longleaf pine savannas and flatwoods and roadsides.
Comments: The flowers of all the skullcaps have one distinctive feature: the calyx has a hump (*scutella*) on the upper side. This is also likened to a *galerum,* a leather skullcap worn by the Romans.

339 **Samson Snakeroot**

Psoralea psoralioides (Walter) Cory
Fabaceae (Pea or Bean Family)
Flowers: May–July
Description: Perennial herb, 1–3' tall; leaves trifoliolate.
Range-Habitat: Common throughout the piedmont and coastal plain; in the coastal plain it occurs in longleaf pine flatwoods and savannas and sandy, dry, open woods, fields, and clearings.
Comments: One of many plants that was believed to act against snake venom.

340 **Colicroot**
Aletris aurea Walter
Liliaceae (Lily Family)
Flowers: Mid-May–July
Description: Perennial herb with thick, short rhizomes; leaves mostly basal, rather narrowly lanceolate; stem leaves greatly reduced; flower stalk 16"–4' tall; perianth covered with small, sticky projections giving a mealy appearance.
Range-Habitat: Primarily a coastal plain species; common in longleaf pine flatwoods and savannas and pocosin borders.
Comments: Same as for *A. farinosa* (plate 329).

341 **Snowy Orchid**
Habenaria nivea (Nuttall) Sprengel
Synonym: *Platanthera nivea* (Nuttall) Luer
Orchidaceae (Orchid Family)
Flowers: May–September
Description: Herbaceous, erect perennial, to 1' tall; leaves two to three, near the base, rigidly suberect, linear-lanceolate, reduced to as many as ten slender, erect bracts.
Range-Habitat: A coastal plain species; infrequent in longleaf pine savannas and moist sites in longleaf pine flatwoods.
Comments: Like many orchids, snowy orchid may cover a site one year, then be absent for several years before again making an appearance. This orchid depends on a more or less constant supply of acid waters for best development. Many populations have been eliminated because of alteration of drainage patterns near savannas.

The common name is derived from the Latin *niveus* meaning "white as snow," in reference to the intense whiteness of the flowers.

342 **White Sabatia; Baby's Breath**
Sabatia difformis (L.) Druce
Gentianaceae (Gentian Family)
Flowers: May–September
Description: Erect, herbaceous perennial to 3' tall, from a short rhizome; branches of inflorescence opposite, forming a flat or convex top.
Range-Habitat: Throughout the coastal plain; common in longleaf pine savannas and wet, longleaf pine flatwoods.
Similar Species: A similar coastal plain species with white flowers occurring in pine savannas is *S. brevifolia* Raf. This annual species has alternate branches of the inflorescence.
Comments: Both of these species are sold in the Charleston marketplace by the "flower ladies" who refer to them as "baby's breath."

343 **Smooth Meadow-beauty**
Rhexia alifanus Walter
Melastomaceae (Meadow-beauty Family)

Flowers: May–September
Description: Perennial herb to 40" tall; stem hairless; fruiting urn has gland-tipped hairs; leaves prominently three-nerved; the eight anthers are prominently curved.
Range-Habitat: A coastal plain species; common throughout in longleaf pine savannas and low, longleaf pine flatwoods and ditches.

344 **Pale Meadow-beauty**
Rhexia mariana L.
Melastomaceae (Meadow-beauty Family)
Flowers: May–October
Description: Branched or unbranched, perennial herb; often colonial from elongate, horizontal roots; stems to 32" tall; petals white to purplish; anthers prominently curved; stems bear gland-tipped hairs.
Range-Habitat: Common throughout the state; in the coastal plain in longleaf pine savannas, roadside ditches, and meadows.
Taxonomy: Several varieties are recognized that are too complex to deal with in this book. The reader may refer to Radford et al. (1968) or Duncan and Foote (1975) for more information.
Comments: The capsules take on a distinctive urn-shaped appearance and become a ready means of identifying the genus. Even in the winter, dead stalks of meadow-beauty can be identified.

345 **Virginia Meadow-beauty; Deergrass**
Rhexia virginica L.
Melastomaceae (Meadow-beauty Family)
Flowers: May–October
Description: Perennial herb, with stems to 40" tall, spongy thickened near the base; stems usually branched, with coarse hairs; roots usually tuberous; stem winged-angled with four, nearly equal faces at midstem; leaves conspicuously ciliate-toothed on margins; anthers curved.
Range-Habitat: Common throughout the state but less so in the piedmont; common in the coastal plain in wet, longleaf pine savannas; edges of swamp-gum ponds; and roadside ditches.
Comments: The young leaves make an excellent salad and have a sweetish and slightly acid taste. The edible tubers are pleasantly nutty.

346 **Stylisma**
Bonamia aquatica (Walter) Gray
Synonym: *Stylisma aquatica* (Walter) Raf.
Convolvulaceae (Morning-glory Family)
Flowers: June–July
Description: Herbaceous, creeping or spreading, perennial vine; corolla pink or purple.

Range-Habitat: A coastal plain species; infrequent in moist, longleaf pine savannas.

347 **False Asphodel**
Tofieldia racemosa (Walter) BSP.
Liliaceae (Lily Family)
Flowers: June–Early August
Description: Perennial herb from a short rhizome; flowering stalk 12–28" tall; basal leaves erect, linear, variable, to 16" long; usually one bractlike leaf inserted somewhere below the middle of the flowering stalk; flowering stalk rough to the touch.
Range-Habitat: A coastal plain species; common in longleaf pine savannas and flatwoods and pocosins.

348 **Savanna Sabatia**
Sabatia campanulata (L.) Torrey
Gentianaceae (Gentian Family)
Flowers: June–August
Description: Perennial herb with short, much-branched rhizome; stems slightly four-angled, 12–28" tall; no basal leaves; stem leaves linear to narrowly lanceolate.
Range-Habitat: A coastal plain species; common in longleaf pine savannas.

349 **Toothache Grass; Orange-grass**
Ctenium aromaticum (Walter) Wood
Synonym: *Campulosus aromaticus* (Walter) Trinius
Poaceae (Grass Family)
Flowers: June–August
Description: Tufted, erect perennial from short rhizomes; stems 2–4' tall; the flowering head is distinctive to *Ctenium;* leaves mostly near the base.
Range-Habitat: A coastal plain species; common in longleaf pine savannas and wet areas in longleaf pine flatwoods.
Comments: The fresh herbage, inflorescence, and rhizome produces a strong citrus aroma when bruised or crushed.

350 **Dwarf Milkwort**
Polygala racemosa Ell.
Polygalaceae (Milkwort Family)
Flowers: June–August
Description: Biennial herb to 12" tall; stems usually one, from a basal cluster of leaves which are frequently absent at flowering; stem leaves reduced; flowers yellow, turning dark green on drying.
Range-Habitat: A common coastal plain species; found in longleaf pine savannas; wet, longleaf pine flatwoods; and pond margins.

351 **Marsh-fleabane**
Pluchea rosea Godfrey
Asteraceae (Aster or Sunflower Family)
Flowers: June–August
Description: Aromatic, perennial herb with a strong, camphor fragrance; 16–44" tall; leaves alternate with sessile, clasping bases.
Range-Habitat: Common coastal plain species; longleaf pine savannas, ditches, pond shores, intermittent ponds, and poorly drained woods.

352 **Pineland Hibiscus; Comfort-root**
Hibiscus aculeatus Walter
Malvaceae (Mallow Family)
Flowers: June–August
Description: Perennial herb to 3' tall; stems spreading-ascending or erect; stems and leaves rough from short, stout hairs; corolla with a purple center, lobes cream-colored initially, fading to yellow, then finally to pink as the petals wither.
Range-Habitat: Throughout the coastal plain; common in longleaf pine savannas and edges of longleaf pine flatwoods.

353 **Redroot**
Lachnanthes caroliniana (Lam.) Dandy
Haemodoraceae (Bloodwort Family)
Flowers: June–Early September
Description: Perennial herb with prominent red rhizome and fibrous roots, both with red juice; flowering plants to 4' tall; leaves mostly basal, linear, rapidly reduced upward.
Range-Habitat: Throughout the coastal plain; common in sandy, longleaf pine savannas, ditches, pocosin borders, and in managed freshwater impoundments along rivers.
Comments: Some sources list redroot as poisonous; Kingsbury (1964), however, disputes this with good evidence. Redroot is a prized food for waterfowl along the coastal rivers; both the seeds and rhizome are used as food. Freshwater impoundments may be managed to encourage the growth of redroot to attract ducks.

354 **Slender Seed-box**
Ludwigia virgata Michaux
Onagraceae (Evening-primrose Family)
Flowers: June–September
Description: Erect, little-branched, perennial herb to 40" tall; leaves alternate, smooth, lanceolate to elliptic, sessile.
Range-Habitat: Throughout the coastal plain; common in longleaf pine savannas.

355 **Crested Fringed-orchid**
Habenaria cristata (Michaux) R. Brown
Synonym: *Platanthera cristata* (Michaux) Lindley
Orchidaceae (Orchid Family)
Flowers: June–September
Description: Perennial, stout herb, 7–35" tall; leafy below, bracteate above; roots numerous, fleshy, tuberous.
Range-Habitat: Throughout the coastal plain; rare in the mountains; common in the coastal plain in longleaf pine savannas and flatwoods, pocosins, and roadsides.
Comments: The common name "crested" comes from the Latin term *cristata* in reference to the exposed fringed tip of the petals.

This orchid represents one noted feature of some members of the orchid family: the great diversity of habitat many species exhibit. Another lesson learned from this species is that some species are found in the mountains and coastal plain but are absent in the piedmont—due primarily to the difference in habitats of the mountains and coastal plain versus the piedmont.

356 **Bitter Mint**
Hyptis alata (Raf.) Shinners
Lamiaceae (Mint Family)
Flowers: June–September
Description: Perennial herb to 40" tall; flowers in dense heads, one head on each stalk; stems four-angled.
Range-Habitat: A coastal plain species; common throughout in longleaf pine savannas, ditches, and wet, longleaf pine flatwoods.

357 **Pine Lily; Catesby's Lily**
Lilium catesbaei Walter
Liliaceae (Lily Family)
Flowers: Mid-June–Mid-September
Description: Perennial herb with unbranched, erect stems from scaly bulbs; stems 20–28" tall; flower solitary, erect; leaves alternate, ascending, or appressed to the stem, the lower about 3.5" long, reduced upward.
Range-Habitat: A coastal plain species; infrequent in longleaf pine savannas and flatwoods.
Comments: This lily is in such a delicate balance with its environment that it does not survive when transplanted into gardens. Its numbers have been greatly reduced for unknown reasons.

358 **Camass**
Zigadenus glaberrimus Michaux
Liliaceae (Lily Family)
Flowers: Late June–Early September
Description: Coarse plant, 32–48" tall, from a thick, hard rhizome; leaves 6–16" long, gradually reduced upward; inner perianth segments bearing two conspicuous glands near base.

Range-Habitat: Throughout the coastal plain; common in longleaf pine savannas and flatwoods and pocosins.

Comments: See comments under *Z. densus* (plate 328) for notes on its poisonous nature.

359 **Polygala**

Polygala mariana Miller

Polygalaceae (Milkwort Family)

Flowers: June–October

Description: Annual herb, to 16" tall; stem smooth and hairless; leaves alternate, linear.

Range-Habitat: Common throughout the coastal plain; grows in longleaf pine savannas and other open, wet habitats.

Comments: The common name of the milkworts comes from an old belief that certain species increase the flow of milk if eaten by cattle. They are not kin to the milkweeds (*Asclepias* ssp.) which derive their name from their milky juice.

360 **Drum-heads**

Polygala cruciata L.

Polygalaceae (Milkwort Family)

Flowers: June–October

Description: Erect, freely branched, annual herb to 14" tall, with wing-angled stems; leaves linear-lanceolate, mostly whorled in threes and fours.

Range-Habitat: Chiefly coastal plain, but also in the mountains; common in the coastal plain in longleaf pine savannas and pocosins.

361 **Yellow Fringeless-orchid; Frog-arrow**

Habenaria integra (Nuttall) Sprengel

Synonym: *Platanthera integra* (Nuttall) Gray ex Beck

Orchidaceae (Orchid Family)

Flowers: July–September

Description: Perennial herb to 2' tall, with smooth stems; roots fleshy, tuberous, swollen near base of stem; lip entire; leaves several below, reduced to bracts above; raceme initially conical but soon becomes cylindrical.

Range-Habitat: A coastal plain orchid; rare in pine savannas and pine flatwoods.

Comments: Frog-arrow is often mistaken for crested fringed-orchid (*H. cristata,* plate 355); it can be identified readily by its entire lip and saffron color.

This orchid is present one year at a site, then absent the next year (or maybe the next few years). No explanation is known for this periodicity.

The name *integra* is Latin for "entire," in reference to the nearly entire, fringeless margin of the lip.

362 **White Fringed-orchid**

Habenaria blephariglottis (Willd.) Hooker

Synonym: *Platanthera blephariglottis* (Willd.) Lindley

Orchidaceae (Orchid Family)
Flowers: July–September
Description: Robust, erect, stout orchid, 1–2.5' tall; perennial from fleshy, tuberous-thickened roots; stems smooth, leafy below, reduced to bracts above; lip fringed.
Range-Habitat: A coastal plain species, scattered throughout; occasional in moist sites in longleaf pine savannas and flatwoods and pocosins.
Comments: Little difference in vegetation occurs between this species and *H. ciliaris* (plate 363) except for flower color. Often they grow in the same site; when this occurs they sometimes produce a hybrid with cream flowers called *Platanthera* x *bicolor.* Several times I have observed this hybrid in the coastal area.

363 **Yellow Fringed-orchid**
Habenaria ciliaris (L.) R. Brown
Synonym: *Platanthera ciliaris* (L.) Lindley
Orchidaceae (Orchid Family)
Flowers: July–September
Description: Erect, robust, perennial herb, 1–2.5' tall; roots fleshy, tuberous-thickened; stem smooth, leafy below, reduced to bracts above; flowers bright yellow or orange; lip fringed.
Range-Habitat: Chiefly a mountains and coastal plain species; common in the coastal plain in longleaf pine savannas and flatwoods, edges of cypress swamps, roadsides, and pocosins.

364 **Blazing Star**
Liatris spicata var. *resinosa* (Nuttall) Gaiser
Asteraceae (Aster or Sunflower Family)
Flowers: July–September
Description: Perennial herb from a thickened, globose rootstock; stems erect, 1–6' tall; leaves numerous, linear, crowded, 1' or more in length at base of plant, decreasing in size upward; bracts below flowering heads usually purple.
Range-Habitat: This variety is strictly a coastal plain plant; it is common in longleaf pine savannas and flatwoods and ditches.
Comments: Blazing star is often cultivated as an ornamental in home gardens.
Taxonomy: This variety is separated from the typical variety in having purple bracts below the flowering heads instead of green; the typical variety occurs in the mountains. Some sources do not separate the two varieties; instead, both are included under *Liatris spicata* (L.) Willd.

365 **Barbara's-buttons**
Marshallia graminifolia (Walter) Small
Asteraceae (Aster or Sunflower Family)
Flowers: Late July–September

Description: Fibrous-rooted, perennial herb; stems 16–32" tall; basal leaves firm and ascending, linear to narrowly elliptic, 2–8" long; stem leaves not strongly different from basal leaves except in size, usually less than 1.2" long; fibrous base of old basal leaves persistent.

Range-Habitat: A coastal plain species; common in longleaf pine savannas and flatwoods and pocosins.

366 **Eryngo**

Eryngium integrifolium Walter

Apiaceae (Parsley Family)

Flowers: August–October

Description: Erect, perennial herb, 8–32" tall; stem solitary, branching above; flowers in subglobose heads.

Range-Habitat: Primarily a coastal plain species; common in longleaf pine savannas and meadows; less common in longleaf pine flatwoods.

Comments: *Eryngium* is one of the few members of the parsley family with flowers in heads rather than umbels. The obvious bracts below each flower head are characteristic of *Eryngium* in general.

367 *Pine–Saw Palmetto Flatwoods*

367 **Saw Palmetto**

Serenoa repens (Bartram) Small

Arecaceae (Palm Family)

Description: Perennial shrub with branched, trailing stems; occasionally a small, branched tree; leaves evergreen, to 3' across, palmately divided into nonfilamentose segments; leaf stalks armed with spines; flowers produced on elongated branches to 3' long; fruit a bluish-black drupe ripe in October–November.

Range-Habitat: Common as a shrub in Beaufort and Jasper counties in pine–saw palmetto flatwoods.

Comments: Saw palmetto often forms dense, almost impenetrable stands in flatwoods. It is highly resistant to fire. The drupes were an important food for Native Americans and for whitetail deer. It can be found in Victoria Bluff Heritage Preserve.

368 **Hairy Wicky**

Kalmia hirsuta Walter

Ericaceae (Heath Family)

Flowers: June–July

Description: Low, evergreen shrub, 6–20" tall; twigs and leaves with stiff hairs; leaves alternate.

Range-Habitat: Chiefly in the southwestern counties of the coastal plain; occasional in pine–saw palmetto flatwoods and longleaf pine flatwoods and savannas.

Comments: In newly opened flowers, the ten stamens are seated in small pockets of the corolla. An insect visiting the mature flowers may cause the stamens to spring out of the pockets, spraying the pollen onto the insect's back which in turn is deposited on the next flower the insect visits, thus effecting cross-pollination.

369 ***Pine–Mixed Hardwood Forests***

THE POND CYPRESS GARDENS

370 ***Pond Cypress Savannas***

371 **Pond Cypress**
Taxodium ascendens Brongn.
Taxodiaceae (Taxodium Family)
Female Cones: All Year
Description: Medium-sized, deciduous tree; leaves needlelike, pressed against the twig, except on seedlings and fast growing shoots where they are two-ranked; male and female cones on the same tree.
Range-Habitat: Confined to the coastal plain; infrequent and scattered throughout in acidic savannas and bogs and nonalluvial swamps.
Comments: When pond cypress grows in wet or frequently flooded conditions, the base tends to grow larger, forming a buttress of lighter and more porous wood. The roots then send up unbranched shoots called knees. In pond cypress, the knees are rounded on top.

Pond cypress wood contains an essential oil that gives it a natural durability which makes it a valuable wood for fence post, shingles, and panelling. Few merchantable stands exists today because reproduction has not kept up with lumbering. It is often used as an ornamental since it grows in upland habitats under cultivation.
Taxonomy: Some sources list pond cypress as a variety of bald cypress: *Taxodium distichum* var. *nutans* (Aiton) Sweet.

372–373 **Yellow Trumpet Pitcher-plant**
Sarracenia flava L.
Sarraceniaceae (Pitcher-plant Family)
Flowers: April–May
Description: Perennial, carnivorous herb from rhizomes; 1.5–3.5' tall; leaves modified to trap insects; may occur in four color forms: (1) pale green to bright yellow in full sun, with a large maroon splotch on the inside of the column from which red veins radiate; (2) bright to deep red color on the external surface of the lid and column, with a weak maroon spot; (3) uniformly golden yellow in full sunlight, with coarse and prominent veins all over and with the interior column spot weak; or (4) no red pigment at all, the mature

pitchers being pale green to yellow; the different forms often grow mixed in the same site; flowers develop in the spring before or with the new leaves; leaves die back during the fall and winter.

Range-Habitat: A common coastal plain species becoming less common from habitat destruction; good populations still exist, however, in the Francis Marion National Forest and other protected lands such as the Santee Coastal Preserve; grows in longleaf pine flatwoods and savannas, openings in pocosins, and pond cypress savannas; in the wetter habitats it grows more robust with the leaves reaching over 3' tall.

Comments: For a general discussion on carnivorous plants, the reader may refer to the essay on carnivorous plants under "Natural History of Selected Groups of Flowering Plants."

374 **Sneezeweed, Bitterweed**

Helenium pinnatifidum (Nuttall) Rydb.
Asteraceae (Aster or Sunflower Family)
Flowers: April–May
Description: Fibrous-rooted, erect perennial, 8–40" tall; commonly one flowering head; basal leaves tufted and persistent; stem leaves few and reduced upward.
Range-Habitat: Primarily a coastal plain species; infrequent in pocosins, pond cypress savannas, and longleaf pine savannas.

375 **Bay Blue-flag**

Iris tridentata Pursh
Iridaceae (Iris Family)
Flowers: Late May–June
Description: Rhizomatous perennial; flowering stalks 12–28" tall, usually unbranched, bearing a single flower.
Range-Habitat: Common in the northeastern coastal plain in longleaf pine savannas, pond cypress savannas, and ditches.

376 **Tall Milkwort**

Polygala cymosa Walter
Polygalaceae (Milkwort Family)
Flowers: May–July
Description: Biennial herb with a solitary, smooth, flowering stem, 16"–4' tall; leaves mostly basal, linear to lanceolate, rapidly reduced upward.
Range-Habitat: Common coastal plain species in shallow waters of pond cypress swamps, pond cypress savannas, and barrow pits.

377 **Sand Weed**

Hypericum fasciculatum Lam.
Hypericaceae (St. John's-wort Family)
Flowers: May–September
Description: Erect shrub, 2.5–5' tall, much-branched above and

spongy-thickened below, with bark peeling in thin sheets.
Range-Habitat: Occasional and scattered throughout the coastal plain in pond cypress savannas, ditches, and wet pinelands.

378 **Giant White-topped Sedge**
Dichromena latifolia Baldwin
Cyperaceae (Sedge Family)
Flowers: May–September
Description: Perennial herb; stems solitary to 4' tall, from elongate rhizomes; leaves appearing mostly basal; inflorescence of crowded clusters of spikes subtended by seven or more widely linear to lanceolate bracts; bracts conspicuously white at base and green at apex.
Range-Habitat: Common coastal plain species in pond cypress savannas, longleaf pine savannas, ditches, depressions, and borders of pocosins.
Comments: The white bracts resemble showy, daisy-like flowers.
Similar Species: *Dichromena colorata* (L.) Hitchcock has five to six bracts and is 20–24" tall with a similar habitat, range, and flowering period.

379 **Awned Meadow-beauty**
Rhexia aristosa Britton
Melastomaceae (Meadow-beauty Family)
Flowers: June–September
Description: Usually branched, perennial herb, to 28" tall; stems four-angled; nodes with stiff hairs; leaves opposite, mostly broadly lanceolate, three-nerved; petals sharp-pointed; capsules (urns) with coarse, yellow bristles on the rim.
Range-Habitat: A rare coastal plain species; pond cypress savannas.

380 **Pipewort; Hatpins; Buttonrods**
Eriocaulon decangulare L.
Eriocaulaceae (Pipewort Family)
Flowers: June–October
Description: Pipewort is conspicuous in flower because of the dense, hard heads of white flowers (buttons) on the tip of leafless stalks; perennial herb with flowering stalks 12–32" tall, finely eight to twelve ridged; leaves mostly basal, linear, with air spaces visible to the naked eye.
Range-Habitat: Common throughout the coastal plain in pond cypress savannas, longleaf pine savannas, sphagnum openings in pocosins, ditches, sandy shores of lakes, and pond cypress swamp forests.
Similar Species: *Eriocaulon compressum* Lam. has heads easily compressed between the fingers and flowering stalks 8–18" tall; occasional in the coastal plain in similar habitats.

381 **Tickseed**
Coreopsis gladiata Walter
Asteraceae (Aster or Sunflower Family)

Flowers: August–October

Description: Short-lived, perennial herb, mostly 20"–4' tall; leaves all alternate, unlobed and variable, with principal ones of most specimens with well-differentiated petioles and blades; rays conspicuously toothed.

Range-Habitat: A coastal plain species infrequent in scattered localities in wet, often acidic habitats such as bogs, ditches, pond cypress savannas, and adjacent longleaf pine savannas.

Similar Species: *Coreopsis falcata* Boynton is a common coastal plain species and grows in similar habitats; however, it flowers May–June and some of its larger leaves have one or two narrow, basal lobes.

382 *Pond Cypress–Swamp-gum Swamp Forests*

383–384 Pond Berry; Southern Spicebush; Jove's Fruit

Lindera melissifolia (Walter) Blume

Synonym: *Benzoin melissifolia* (Walter) Nees

Lauraceae (Laurel Family)

Flowers: March–April

Fruits: August–September

Description: Deciduous, aromatic shrub, 3–4' tall; colonial from rhizomes; leaves entire, alternate; male and female flowers on separate plants; flowers develop before leaves.

Range-Habitat: Rare outer coastal plain species known only in Berkeley and Beaufort counties; along the margins of lime sinks and in wet depressions in pine flatwoods.

Comments: Pond berry is designated as a Federally Endangered Species. The largest concentration of pond berry throughout its range occurs in the Honey Hill region of the Francis Marion National Forest in Berkeley County. The numerous lime sinks in this area provide critical habitat for this rare shrub. Efforts are under way to have this area designated a Research Natural Area.

385 Climbing Fetterbush; Climbing Heath

Pieris phillyreifolia (Hooker) DC.

Synonym: *Ampelothamnus phillyreifolius* (Hooker) Small

Eriaceae (Heath Family)

Flowers: April–May

Description: Shrub or woody vine; as a vine, unique in climbing in crevices of bark or beneath outer bark of *Taxodium ascendens;* as a shrub, it grows on cypress knees or stumps; lateral branches that project from under the bark produce flowers in spring.

Range-Habitat: Known in only four sites in South Carolina, all from the Francis Marion National Forest in Charleston County. Duncan (1975) states it also climbs species of *Pinus,* but rarely. All specimens found in the Francis Marion National Forest climb pond cypress.

386 **Dahoon Holly; Cassena**
Ilex cassine L.
Aquifoliaceae (Holly Family)
Fruits: October–November
Description: Evergreen, small- to medium-sized tree, rarely exceeding 25' tall; leaves simple, obovate, oblanceolate, to narrowly lanceolate, 1.5–4" long and over .25" wide; male and female flowers on separate trees in May–June; fruit berrylike, red or sometimes yellow, persisting until spring.
Range-Habitat: Chiefly outer coastal plain; occasional in upland pond cypress–swamp-gum sites and pocosins.
Comments: Dahoon is often grown as an ornamental for its attractive red berries and evergreen leaves. The fruits are an important food for overwintering songbirds.

387 **Myrtle Dahoon; Myrtle-leaved Holly**
Ilex myrtifolia (Walter) Sargent
Aquifoliaceae (Holly Family)
Fruits: October–November
Description: Small tree or more often a shrub; evergreen, with narrow, stiff leaves, less than 1.5" long and less than .25" wide; male and female flowers on separate plants; fruit berrylike, red, persisting until spring.
Range-Habitat: Chiefly coastal plain; occasional in upland pond cypress–swamp-gum swamps and pocosins.
Taxonomy: Myrtle dahoon and cassena are treated by some manuals as varieties of *I. cassine* and listed as *Ilex cassine* var. *cassine* and *Ilex cassine* var. *myrtifolia.* The differences between the two are based on leaf shapes as described above.

THE RUDERAL GARDENS

388 ***Ruderal Gardens***

389 **Henbit**
Lamium amplexicaule L.
Lamiaceae (Mint Family)
Flowers: Mid-Winter–May
Description: Winter annual to 14" tall; stems soft, four-angled, with freely branched ends; flowering stem ends erect; leaves roundish to ovate, opposite, often shallowly three-lobed; flower clusters subtended by a pair of sessile, leaflike bracts, ascending or horizontal.
Range-Habitat: Naturalized and common throughout the state; in the coastal plain in pastures, abandoned fields, lawns, gardens, and almost any other disturbed sites.
Comments: Henbit is naturalized from Europe. The young plants have been used as a potherb in the United States and in Japan. Kingsbury (1964) reports it caused "staggers" in sheep, cattle, and horses. Eating large quantities is not recommended.

390 **Chickweed**
Stellaria media (L.) Cyrillo
Caryophyllaceae (Pink Family)
Flowers: January–May
Description: Herbaceous, prostrate, or creeping annual, 3–8" tall; weak-stemmed and much-branched; stems with hairs in lines; leaves opposite; five petals, each deeply cleft.
Range-Habitat: Common throughout the state; in the coastal plain in lawns, fields, gardens, meadows, and roadsides.
Comments: Chickweed is naturalized from Eurasia. It has long been used as a potherb; only the young, tender, growing tips are used. It is a favorite food of chickens and wild birds.

391 **Common Blue Violet**
Viola papilionacea Pursh
Violaceae (Violet Family)
Flowers: February–May
Description: Perennial herb with elongate, stocky rhizomes; leaves scalloped; sporadically flowering in the fall; self-pollinating flowers near the ground that do not open produce large quantities of seeds.
Range-Habitat: Common throughout the state; in the coastal plain along roads, in hedge rows, lawns, pastures, and alluvial woods.
Comments: Common blue violet is perhaps the most common violet in the Southeast. Like all violets, its leaves are high in vitamins A and C and can be used in salads or as a potherb. Its flowers can be made into candies and jelly. Francis Porcher (1869) relates that the leaves were made into soup during the Civil War.

392 **Common Dandelion**
Taraxacum officinale Wiggers
Asteraceae (Aster or Sunflower Family)
Flowers: Late February–June
Description: Perennial herb; root thick, deep, bitter; milky juice present; leaves basal, deeply and irregularly lobed and toothed; flowering stalks hollow, erect, 2–18" tall.
Range-Habitat: Naturalized throughout the state; common in the coastal plain in lawns, roadsides, pastures, vacant lots, and fallow fields.
Comments: Common dandelion is a native of Eurasia. In the Old and New World, the leaves have been and are used as a potherb, and the ground roots can be used to make a palatable, bitter drink. Also the roots can be cooked for food. The specific epithet also reveals other uses of this versatile plant (*officinale* means "of the shops"). Its medicinal uses range from alleviating symptoms of arthritis to serving as a laxative. A strong wine can be made from the flowers and leaves.

393 **White Clover**
Trifolium repens L.
Fabaceae (Pea or Bean Family)

Flowers: February–September

Description: Creeping, perennial herb, 4–10" high; roots at the nodes; often forming large masses; most plants usually die back in hot summers; leaves palmately three-lobed.

Range-Habitat: Naturalized from Europe and widespread and common throughout the state; in the coastal plain in lawns, roadsides, fields, and abandoned lots.

394 **Mock Strawberry; Indian Strawberry**

Duchesnea indica (Andrz.) Focke

Rosaceae (Rose Family)

Flowers: February–Frost

Description: Low, trailing, perennial herb with stolons; leaves trifoliolate; fruit an aggregate of achenes embedded in a red, fleshy receptacle.

Range-Habitat: Naturalized throughout the state; common in the coastal plain in lawns, pastures, roadsides, and open woods.

Comments: According to some sources (for example, Strausbaugh and Core, 1977), this species was introduced from India. Although the fruits appear edible, they are flat and tasteless. It is not a true strawberry which would belong to the genus *Fragaria*.

395 **Dewberry**

Rubus trivialis Michaux

Rosaceae (Rose Family)

Flowers: March–April

Description: Trailing vine, the stems rooting at the apex to form extensive colonies; stems with stiff, gland-tipped hairs; stem prickles short and curved; leaves compound with five leaflets on the new shoots, three leaflets on the older, flowering stems; berry ripe late April–May.

Range-Habitat: Lower piedmont and coastal plain; common in the coastal plain along roads and railroads, old fields, dunes, powerlines, and abandoned house sites.

Comments: Dewberries (and blackberries) have been used throughout history by humans in a variety of ways. Historically for the Lowcountry, Francis Porcher (1869) relates that the root was a valuable astringent, and a decoction from it easily checks diarrhea. A laxative was made from the fruits which worked because of the mechanical irritation of the seeds. From the fruits could be made wine, syrup, cordials, jelly, and jam. The reader may refer to the books on edible plants in the bibliography for a more complete list of the uses of dewberries (and blackberries).

Dewberries and related blackberries are among the most important summer foods for songbirds and many mammals.

396 **Golden Corydalis**
Corydalis micrantha (Engelm.) Gray
Fumariaceae (Fumatory Family)
Flowers: March–April
Description: Winter annual or biennial herb; stems erect or creeping, branching from the base, 4–16" tall; racemes above the leaves; upper petal with pronounced spur.
Range-Habitat: Primarily an outer coastal plain species; occasional on sandy roadsides and fallow, sandy fields.

397 **Cherokee Rose**
Rosa laevigata Michaux
Rosaceae (Rose Family)
Flowers: Late March–April
Description: Robust, high-climbing, evergreen vine; leaves trifoliolate; leaflets with prickles on larger veins; thorns on stems curved, flattened, with a broad base; hip red, maturing September–October.
Range-Habitat: Primarily naturalized in the coastal plain; occasional around abandoned home sites, roadways, and low woods.
Comments: Cherokee rose is native to China and Japan; it was introduced early into the South and became naturalized. It was actually first described from plants collected in America.

398 **False-garlic**
Allium bivalve (L.) Kuntze
Synonym: *Nothoscordum bivalve* (L.) Britton
Liliaceae (Lily Family)
Flowers: March–May; September–October
Description: Perennial herb from a small, onionlike bulb; bulb without onionlike odor; flowering stalks 6–18" tall; leaves basal, linear, flat.
Range-Habitat: Common in the coastal plain and lower piedmont; rare in inner piedmont; in the coastal plain it occurs in fields, lawns, roadsides, and pastures.

399 **Sourgrass; Wild Sorrel**
Rumex hastatulus Baldwin ex Ell.
Polygonaceae (Buckwheat Family)
Flowers: March–May
Description: Annual or short-lived perennial to 4' tall; leaves arrow-shaped with spreading, basal lobes; stems single or in large clumps; male and female flowers on separate plants.
Range-Habitat: Common throughout the state; in the coastal plain found in sandy fields, roadsides, vacant lots, lawns, and sandy, open woods.
Comments: A native of the Southeast, sourgrass often grows in

association with blue toadflax in fallow fields where these two species form a colorful mix of red and blue. The stem of sourgrass has an acid taste and is often chewed as a trail nibble. Eating large amounts, however, can result in poisoning because of the oxalates present.

400 **Toadflax**

Linaria canadensis (L.) Dumont
Scrophulariaceae (Figwort Family)
Flowers: March–May
Description: Winter annual or biennial herb; erect, flowering stems to 30" tall; stems slender with linear, alternate leaves; numerous prostrate stems with opposite leaves radiate from base of upright stem; conspicuous spur projects down from the corolla.
Range-Habitat: Common throughout the state; in the coastal plain in fallow fields, roadsides, lawns, pastures, and vacant lots.
Comments: Toadflax and sourgrass often grow together in fallow fields where they form a colorful mix of red and blue. Both species are native but grow more as weeds.

401 **Wild-pansy; Johnny-jump-up**

Viola rafinesquii Greene
Violaceae (Violet Family)
Flowers: March–May
Description: Winter annual, about 4–6" tall; stems simple or branched above; lower three petals with purple veins.
Range-Habitat: Common in the mountains and piedmont; occasional in the coastal plain along roads, in lawns, and pastures.
Comments: Botanists disagree whether this violet is native or introduced. It often forms dense but ephemeral populations, especially along roadsides.

402 **Carolina Cranesbill**

Geranium carolinianum L.
Geraniaceae (Geranium Family)
Flowers: March–June
Description: A winter annual, 1–2' tall, from a small taproot; stems in a rosette, prostrate to ascending; leaves palmately lobed; petals pale pink to nearly white.
Range-Habitat: A common weed over the entire state; in the coastal plain in lawns, roadsides, gardens, fields, pastures, and vacant lots.

403 **Yellow Thistle**

Carduus spinosissimus Walter
Synonym: *Cirsium horridulum* Michaux
Asteraceae (Aster or Sunflower Family)
Flowers: Late March–Early June
Description: Biennial herb to 5' tall; stems covered with cobweb-like hairs; leaves spiny on the margin, pinnately lobed,

stalkless, and clasping the stem; flowering heads subtended by a series of narrow, spiny-toothed leaves; corolla light yellow to purple.
Range-Habitat: Common throughout the piedmont and coastal plain; in the coastal plain it occurs in longleaf pine savannas, roadsides, fields, and meadows.
Taxonomy: Two forms occur, yellow flowered and purple flowered; the former restricted to the coastal plain, the latter grows throughout the state.

404 **Dwarf Dandelion**
Krigia virginica (L.) Willd.
Asteraceae (Aster or Sunflower Family)
Flowers: Late March–Early June
Description: Slender annual to 16" tall; juice milky; flowering stems one to several; leaves generally basal, toothed, or lobed.
Range-Habitat: Common throughout the state; in the coastal plain common in lawns, roadsides, fields, pastures, and open woods.

405 **Prickly Sow-thistle**
Sonchus asper (L.) Hill
Asteraceae (Aster or Sunflower Family)
Flowers: March–July
Description: Herbaceous, winter annual, 1–6' tall; stems smooth, erect; leaves alternate, lobed or undivided, prickly edged, base rounded with an ear-shaped lobe; white juice present.
Range-Habitat: Common throughout the state; in the coastal plain along roadsides, in fields, vacant lots, lawns, and meadows.
Similar Species: *S. oleraceus* L., common sow-thistle, is similar but has a leaf base shaped like an arrowhead; common and in the same habitat.
Comments: Prickly sow-thistle is naturalized from the Old World; the fleshy leaves were used as a potherb.

406 **Cut-leaved Evening-primrose**
Oenothera laciniata Hill
Onagraceae (Evening-primrose Family)
Flowers: March–July
Description: Hairy, biennial herb; stems usually creeping; leaves nearly entire to irregularly cut; flowers reddish brown or yellow.
Range-Habitat: Common throughout the state; in the coastal plain in fallow fields, lawns, vacant lots, roadsides, and gardens.

407 **Moss Verbena**
Verbena tenuisecta Briquet
Verbenaceae (Vervain Family)

Flowers: March–Frost
Description: Prostrate to creeping, perennial herb, rooting at the nodes; stems branch abundantly, forming mats; leaves opposite, divided into numerous linear segments; corolla lavender, purple, pink, or white.
Range-Habitat: Common throughout the piedmont and coastal plain; in the coastal plain along roads, in pastures, fields, lawns, and vacant lots.
Comments: Moss verbena is a native of tropical America and is now widely naturalized. It is especially abundant along roadsides and does not seem to be adversely affected by repeated mowing.

408 **Lyre-leaved Sage**
Salvia lyrata L.
Lamiaceae (Mint Family)
Flowers: April–May
Description: Perennial herb, with one, or rarely two, pairs of stem leaves; stems four-angled, 12–32" tall; five to fifteen basal leaves, pinnately lobed or dissected or unlobed, often lyre-shaped.
Range-Habitat: Common throughout the state; in the coastal plain occasional in open oak-hickory forests; very common along roadsides, yards, and meadows.
Comments: An interesting pollen-spreading mechanism is seen in the flowers. When a bee lands on the exposed lower lip, it tips the two stamens downward which douse the bee with pollen.

409 **Prickly Poppy**
Argemone albiflora Hornemann
Papaveraceae (Poppy Family)
Flowers: April–May
Description: Annual, prickly herb, 1–3' tall; juice white but turns yellow after drying; leaves irregularly and coarsely pinnately parted.
Range-Habitat: Chiefly a coastal plain species infrequent in a variety of disturbed places: roadsides, railroads, abandoned city lots, and open fields; a naturalized species.
Similar Species: *A. mexicana* L., yellow prickly poppy, has yellow flowers. It was introduced from tropical America, and its vegetative parts and seeds are poisonous. It occurs in the same habitats as prickly poppy and is rare in the coastal plain

410 **China-berry; Pride-of-India**
Melia azedarach L.
Meliaceae (Mahogany Family)
Flowers: April–May
Description: Small- to medium-sized tree; stems aromatic; leaves alternate, bipinnately compound; inflorescence a panicle; fruits yellow, ripe September–October.
Range-Habitat: Naturalized and common throughout the state,

except at high altitudes; in the coastal plain along woodland borders, fence rows, vacant lots, and persisting around abandoned homes.

Comments: A native of southwest Asia, China-berry was introduced into America by André Michaux to his garden in Charleston. From this and other gardens, it escaped and became naturalized.

China-berry quickly came to have many uses for plantation slaves and colonists. It was used as a vermifuge to expel worms from the body; broken branches were placed in a house to keep out fleas. Its use as a fleabane is based on chemicals in the wood repulsive to insects.

Both the green and ripe fruits of China-berry are poisonous, although poisoning is rare because of the bitter taste of the fruits.

411 Black Locust

Robinia pseudo-acacia L.
Fabaceae (Pea or Bean Family)
Flowers: April–June

Description: Medium-sized, short-lived tree, 40–60' tall; often suckering extensively from roots; leaves alternate, odd-pinnately compound; petals white except for yellow patch; flowers in loose hanging clusters, fragrant; pair of sharp spines at the base of the leaf stalks.

Range-Habitat: Native to mountains and piedmont; naturalized and occasional in the coastal plain along fence rows, in thickets and woods, and along roadsides.

Comments: The natural range of this tree is the central Appalachian and Ozark mountains. It has been extensively cultivated for its durable wood and ornamental value. Specimens in the Lowcountry have escaped from cultivation. Its wood was used for shipbuilding and exported to England for that purpose. Today it is used for fence posts, railroad ties, and craft items.

Black locust is a nitrogen fixer and can grow in mineral-poor soils; for this reason, it is used for erosion control and on reclamation sites.

412 Moth Mullein

Verbascum virgatum Stokes
Scrophulariaceae (Figwort Family)
Flowers: April–June

Description: Erect annual or biennial to 4' tall, with small, scattered, glandular hairs; stems simple, rarely with one to three secondary branches; stem leaves alternate, sessile, hairy.

Range-Habitat: Naturalized from Eurasia and restricted to the coastal plain where it is frequent along roadsides, railroad tracks, abandoned city lots, and other disturbed areas.

Similar Species: *V. blattaria* L., also introduced from Eurasia, has one flower at the same point of attachment, while *V. virgatum* has two to three flowers at the same point of attachment on the stem. The habitat of both is similar.

Comments: The common name refers to the resemblance between the fuzzy stamens and the antennae of a moth.

413 **Bigflower Vetch**

Vicia grandiflora Scopoli

Fabaceae (Pea or Bean Family)

Flowers: April–June

Description: Weakly climbing annual to 2' tall; leaves alternate, pinnately compound with six to fourteen leaflets and a weak, branched, terminal tendril; petals pale yellow or often suffused with purple.

Range-Habitat: Naturalized from Europe and generally throughout the piedmont and coastal plain; occasional in the coastal plain in fields, abandoned city lots, roadsides, and other open, disturbed sites.

414 **Venus' Looking-glass**

Specularia perfoliata (L.) DC.

Synonym: *Triodanis perfoliata* (L.) Nieuw.

Campanulaceae (Bluebell Family)

Flowers: April–June

Description: Herbaceous, winter annual, 8–40" tall; stems erect, freely branched at base, unbranched above; leaves alternate, sessile, and clasping; flowers sessile in leaf axils; lower flowers do not open and are self-pollinating.

Range-Habitat: Common throughout the state; in the coastal plain common in fields and gardens, roadsides, abandoned lots, and lawns.

Similar Species: *S. biflora* (R. & P.) Greene is similar and usually occurs mixed with *S. perfoliata.* The easiest field character is probably the capsule. In *perfoliata* the pores of the capsule are at or below the middle, while in *biflora* the pores are near the top.

415 **False-dandelion**

Pyrrhopappus carolinianus (Walter) DC.

Asteraceae (Aster or Sunflower)

Flowers: April–June

Description: Milky-juiced annual or short-lived perennial from well-developed taproot; to 4' tall; basal leaves pinnately lobed or dissected to merely toothed; upper leaves reduced; heads several or often solitary.

Range-Habitat: Common throughout the state; in the coastal plain common in pastures, roadsides, fallow and cultivated fields, lawns, and meadows.

416 **Crimson Clover**

Trifolium incarnatum L.

Fabaceae (Pea or Bean Family)

Flowers: April–June

Description: Erect, annual herb, 8–16" tall.
Range-Habitat: Chiefly piedmont and coastal plain; in the coastal plain common along roadsides and fields.
Comments: Crimson clover is naturalized from Eurasia. In fields it is planted as fodder or green manure and is planted by highway departments along roads as a manure, soil binder, and as an attractive flowering plant. The fibrous calyx from overripe flowers may be dangerous to horses because it may become impacted in their digestive tract.

Crimson clover is the only clover in the Lowcountry with red flowers.

417 **Japanese Honeysuckle**
Lonicera japonica Thunberg
Caprifoliaceae (Honeysuckle Family)
Flowers: April–June
Description: Left to right twining, woody vine with opposite, evergreen leaves; often climbing to 30' or more; berry black, maturing in August–October; flowers sporadically into September.
Range-Habitat: Common throughout the state; in the coastal plain found in almost any disturbed habitat including woodlands, fields, fence rows, thickets, abandoned buildings, and along railroad banks.
Comments: Honeysuckle is widely naturalized from Japan and very quickly invades any disturbed openings in native woodlands, sometimes replacing the native flora. It spreads rapidly by seeds carried by birds.

A sweet nectar can be sucked from the base of the corolla.

418 **Ohio Spiderwort**
Tradescantia ohiensis Raf.
Commelinaceae (Spiderwort Family)
Flowers: April–July
Description: Rhizomatous, perennial herb, 8–32' tall; often forming large clumps with numerous stems bearing open umbels; stems leafy; leaves linear.
Range-Habitat: Scattered throughout the state; occasional in the coastal plain in sandy roadsides, dry woods, stable dunes, and fence rows.
Comments: Ohio spiderwort is native to North America; specimens in South Carolina probably escaped from cultivation.

419 **Chinese Wisteria**
Wisteria sinensis (Sims) Sweet
Fabaceae (Pea or Bean Family)
Flowers: April–July
Description: High climbing, woody vine to 65' long; leaves alternate, odd-pinnately compound, with nine or eleven leaflets; racemes terminal and hanging down; legumes

densely velvety hairy, maturing July–November.

Range-Habitat: Naturalized in the piedmont and coastal plain; common in the coastal plain around abandoned house sites, open woods, and roadsides.

Comments: Chinese wisteria is a native of China and was brought to the New World to adorn gardens. The seeds are reported to be poisonous.

420 **Salt Cedar; Tamarisk**

Tamarix gallica L.

Tamaricaceae (Tamarisk Family)

Flowers: April–July

Description: Shrub with evergreen, scalelike, green leaves; branches flexible; fruit a capsule of many seeds; tip of seeds with a tuft of hairs.

Range-Habitat: Primarily an outer coastal plain species; occasional and local in sandy roadsides, dredged soil disposal sites, old fields, and brackish areas.

Comments: Salt cedar is naturalized from Europe. The common name, salt cedar, refers to is coastal distribution and its small, leafy twigs resembling forms of southern red cedar. Salt cedar is a flowering plant (angiosperm), not a conifer (gymnosperm) like southern red cedar.

421 **Annual Phlox**

Phlox drummondii Hooker

Polemoniaceae (Phlox Family)

Flowers: April–July

Description: Erect, herbaceous annual, 4–28" tall; stems glandular-hairy; lowermost leaves opposite, others alternate; flowers rose-red, pink, white, or variegated depending on the cultivar.

Range-Habitat: Chiefly a coastal plain species; common around abandoned house sites, lawns, sandy roadsides, stable dune areas, and meadows.

Comments: This phlox is a native of Texas; it is widely cultivated and has escaped and become naturalized. There are numerous cultivated forms; the different forms are often found growing together, especially around abandoned house sites.

422 **Black Medic**

Medicago lupulina L.

Fabaceae (Pea or Bean Family)

Flowers: April–August

Description: Annual herb, branches 4–16" long, spreading; leaves trifoliolate; flowers small, in dense, axillary heads; legumes curved, black when drying.

Range-Habitat: Common and throughout the state; in the coastal plain along roadsides, in lawns, gardens, and fields.

Comments: Black medic is naturalized from Europe; it is often planted along roads where it forms a bank of yellow in the spring.

423 **Heliotrope**
Heliotropium amplexicaule Vahl
Boraginaceae (Borage Family)
Flowers: April–September
Description: Perennial herb from a strong root; stems several and spreading, creeping, or ascending to 20" tall; plant glandular-hairy; leaves alternate.
Range-Habitat: A piedmont and coastal plain species; in the coastal plain occasional along city roadsides, cultivated fields, and abandoned lots.

424 **Sour Clover**
Melilotus indica (L.) All.
Fabaceae (Pea or Bean Family)
Flowers: April–October
Description: Annual, spreading or ascending, 4–20" tall; leaves alternate, trifoliolate, the terminal leaflet stalked.
Range-Habitat: Widespread throughout the piedmont and coastal plain; common in the coastal plain along roadsides, abandoned city lots, fences, back beaches, and fields.
Comments: A naturalized species introduced for soil improvement.
Similar Species: White sweet clover (*M. alba* Desr.) has white flowers and occurs in the same habitats as sour clover.

425 **Heal-all; Self-heal**
Prunella vulgaris L.
Lamiaceae (Mint Family)
Flowers: April–Frost
Description: Perennial herb to 6–12" tall, with short branches anywhere below the central inflorescence; stems four-angled, leaves opposite; flowers in globose spikes, but spikes becoming shaped like a cylinder as fruits mature.
Range-Habitat: Common throughout the state; in the coastal plain common along roadsides, in lawns, fields, yards, pastures, and gardens.
Comments: As one common names suggests, this plant has been used for a variety of folk remedies in the belief it could cure all ailments. For example, an infusion of the flowers and leaves has been used as a gargle for sore throats.

The typical form of heal-all was naturalized from Europe; American heal-all is var. *lanceolata* and is listed as growing in the same area as the typical variety.

With repeated mowing or grazing, the plants become small, depressed, matted, and flower when only 2" tall.

426 **English Plantain**
Plantago lanceolata L.
Plantaginaceae (Plantain Family)
Flowers: April–Frost
Description: Stemless, perennial herb with a basal cluster of long, narrow, strongly ribbed leaves; flowering stalks solid,

five-angled, 4–20" tall; flowers in dense, cylindrical heads.
Range-Habitat: Common throughout the state; in the coastal plain in lawns, vacant lots, pastures, fields, roadsides, and along railroads.
Comments: English plantain is naturalized from Europe; it is often a troublesome weed in lawns.

427 **Yarrow; Milfoil**
Achillea millefolium L.
Asteraceae (Aster or Sunflower Family)
Flowers: April–Frost
Description: Rhizomatous, perennial herb; stems erect, 1–3' tall; leaves basal and on the stem, finely two or three times pinnately dissected, giving a feathery look; overwinters as a basal rosette.
Range-Habitat: Throughout the state; common in a variety of disturbed places such as pastures, old fields, lawns, roadsides, and meadows.
Comments: Yarrow is naturalized from Asia and Europe. Because of its reputed medicinal value, it was brought with the colonists and planted in their gardens. It was used as a styptic, and a tea made from the leaves was used as a febrifuge and to cure stomach disorders. Today yarrow is valued in flower gardens.

428 **Shepherd's Needle**
Bidens pilosa L.
Asteraceae (Aster or Sunflower Family)
Flowers: April–Frost
Description: Annual herb, to 6' tall; stems erect or creeping at the base; leaves opposite, mostly pinnately three- to five-parted or compound; achenes spindle-shaped with downward-pointing barbs that aid in dispersal by animals.
Range-Habitat: Scattered and infrequent in the coastal plain in disturbed places such as around buildings, roadsides, lawns, old fields, and abandoned city lots.

429 **Popcorn Tree; Chinese Tallow Tree**
Sapium sebiferum (L.) Roxb.
Euphorbiaceae (Spurge Family)
Flowers: May–June
Description: Small- to medium-sized tree, to 50' or taller; easily spreading; leaves alternate, with a pair of glands near base of blade; flowers produced in long, slender spikes, male flowers above, female flowers near the base; fruits maturing August–November; capsule walls fall away, exposing the white seeds.
Range-Habitat: Chiefly in the outer coastal plain; common in maritime forests, edge of fields, disturbed areas in coastal cities, barnyards, and ditch banks.
Comments: Popcorn tree is native to China and Japan but was introduced into the colonies as early as the 1700s as an ornamental or shade tree. It quickly became naturalized. The waxy coating on the seeds was used to make soap and

candles by the Chinese. All parts contain a poisonous milky juice. The common name, popcorn tree, alludes to the cluster of white seeds which give the look of popcorn. In the Lowcountry the seeds are used in Christmas decorations.

430 **Calliopsis**
Coreopsis basalis (Dietrich) Blake
Asteraceae (Aster or Sunflower Family)
Flowers: May–July
Description: Tap-rooted annual or biennial herb; stems erect, freely branched; stem leaves and often basal leaves, pinnately to bipinnately dissected; rays yellow with characteristic red-brown mark at base.
Range-Habitat: Throughout the coastal plain; common along roadsides and in sandy fields.
Comments: This species is native to Texas and southwest Louisiana; locally introduced into gardens, it escaped and is now naturalized.
Similar Species: *C. tinctoria* is similar and also has a red-brown mark at the base of the ray; but *C. tinctoria* has shorter outer bracts below the flowering head, up to half as long as the inner bracts; *C. basalis* has outer bracts more or less elongate, half to fully as long as the inner bracts. *C. tinctoria* is also a western introduction.

431 **Man-of-the-earth; Man-root**
Ipomoea pandurata (L.) G. F. W. Meyer
Convolvulaceae (Morning-glory Family)
Flowers: May–July
Description: Perennial, trailing vine from a deep, vertical, enlarged root up to 30 lbs.; corolla white, always with a purple center.
Range-Habitat: Common throughout the state; in the coastal plain along roadsides, lawns, fence rows, sandy open woods, and fallow fields.
Comments: Several sources state the root was used by the Native Americans but only after long roasting. It is interesting to note that man-of-the-earth is related to sweet potato (*Ipomoea batatas*).

432 **Black-eyed Susan**
Rudbeckia hirta L.
Asteraceae (Aster or Sunflower Family)
Flowers: May–July
Description: Tap-rooted annual to more often biennial or fibrous-rooted perennial; 12–40" tall; leaves and stems very rough and bristly hairy; leaves alternate.
Range-Habitat: Throughout and common in the state; in the coastal plain along roadsides and in pastures, fields, and meadows.
Comments: Black-eyed Susan is native to the prairies of the Midwest but has spread eastward. It has been used as a dye source and to treat skin infections.

433 **Maypops; Passion-flower**

Passiflora incarnata L.
Passifloraceae (Passion-flower Family)
Flowers: May–July
Description: Tendril-bearing, perennial, herbaceous vine; either creeping and rooting at the nodes, or climbing, to 6' long; leaves deeply three-lobed; fruit fleshy, yellow when ripe in July–October.
Range-Habitat: Common throughout the state; in the coastal plain along roadsides, in fallow fields, hedge rows and fences, vacant lots and thickets.
Comments: Passion-flower is a native vine that is now weedy and often cultivated as an ornamental. The common name, passion-flower, comes from resemblance of the floral parts to the story of the Passion of Christ: the styles were fancied to resemble nails; the five stamens, the wounds Jesus received; the purplish corona, the bloody crown; the ten perianth parts, the ten disciples (Peter and Judas being absent); the coiled tendrils, the whips for scourging; the pistil signifies the column where Christ was scourged; and the flower in the background of dull, green leaves represents Christ in the hands of His enemies. And interestingly, the flower's life is generally three days.

The fruit is edible raw but is more esteemed when made into jelly.
Similar Species: *P. lutea* L. has similarly shaped flowers but smaller and greenish yellow; it flowers in June–September and is common in the coastal plain in thickets and woodlands.

434 **Bull Nettle; Horse Nettle**

Solanum carolinense L.
Solanaceae (Nightshade Family)
Flowers: May–July
Description: Erect, weakly branched, perennial herb, 1–3' tall; stems and underside of leaves with sharp prickles; leaves coarsely lobed, both surfaces with star-shaped hairs; corolla light purple to white, with yellow center; berry yellow when ripe.
Range-Habitat: Common throughout the state; coastal plain habitats include roadsides, fallow and cultivated fields, farm lots, abandoned house sites, and lawns.
Comments: The berries are poisonous and Kingsbury (1964) reports one case of a child dying from eating the berries. Poisoning in cattle and deer have been reported. This is another native species that has become weedy.

435 **Mimosa; Silk Tree**

Albizia julibrissin Durazzini
Fabaceae (Pea or Bean Family)
Flowers: May–August
Description: Small tree with large, bipinnately compound and featherlike leaves; to about 50' tall; flowers are

clustered in fluffy, pink heads.

Range-Habitat: Common throughout the state; in the coastal plain it occurs in a variety of disturbed sites such as roadsides, woodland borders, and abandoned house sites.

Comments: Silk tree is an introduced tree, cultivated for its wide-spreading crown, showy flowers, and graceful leaves. It is native to Asia and was introduced by André Michaux.

436 **Smooth Vetch**

Vicia dasycarpa Tenore

Fabaceae (Pea or Ben Family)

Flowers: May–September

Description: Annual or rarely perennial herb, trailing or climbing by tendrils; leaves alternate, pinnately compound with fourteen to twenty leaflets; terminal leaflet modified into a branched tendril; racemes with flowers borne on one side; calyx bulging at base, the flower stalk appearing lateral.

Range-Habitat: Naturalized from Europe and common throughout the state; in the coastal plain in fallow fields, roadsides, fence rows, and other open, disturbed sites.

437 **Queen Anne's Lace; Wild Carrot**

Daucus carota L.

Apiaceae (Parsley Family)

Flowers: May–September

Description: Biennial herb, 1–3' tall, with long, slender taproot; leaves bipinnately compound; fruits with bristles.

Range-Habitat: Common in the mountains, piedmont, and inner coastal plain; occasional in outer coastal plain; most often along roadsides but also in fallow fields, fence rows, and abandoned lots.

Comments: Wild carrot is naturalized from Europe and is widely spread as a weed in North America. For centuries the root was eaten in Europe and early America; however, the root was bitter and stringy and needed long cooking. As is often the case, a wild plant is recognized as having important qualities (here vitamin A), and a cultivar is developed from it, in this case the commercial carrot, *D. carota* var. *sativa.*

438 **Spiny Nightshade**

Solanum sisymbriifolium Lam.

Solanaceae (Nightshade Family)

Flowers: May–September

Description: Coarse, sticky annual herb to 3' tall; stem leaves, flower stalks, and calyx armed with flat, orange-yellow prickles to .5" long; leaves pinnately parted; corolla white to purple; mature berries red, ripe September–October.

Range-Habitat: Occasional in the outer coastal plain in various disturbed places.

Comments: Both Radford et al. (1968) and Duncan and Duncan

(1987) list this species as an annual. In the Charleston area the stems have been observed to over-winter and produce leaves and flowers the following growing season; thus, under some conditions it may be a perennial (or biennial).

439 South American Vervain

Verbena bonariensis L.
Verbenaceae (Vervain Family)
Flowers: May–October
Description: Erect, perennial herb, 3–8' tall; stems four-angled, rough on the angles; leaves opposite, simple, sessile with clasping bases.
Range-Habitat: A common introduction from South America, naturalized throughout the coastal plain and piedmont; common along roadsides, in fields, vacant lots, along fence rows, and ditches.
Similar Species: *V. brasiliensis* Vellozo is similar but has acutely narrowed leaf bases rather than clasping; habitat and range similar.

440 Common Purslane; Pusley

Portulaca oleracea L.
Portulacaceae (Purslane Family)
Flowers: May–October
Description: Fleshy annual with creeping to prostrate stems; leaves alternate and opposite; flowers open only in the sun; fruit opening by a lid just below the middle in a even, circular line.
Range-Habitat: Throughout the state; occasional in gardens, lawns, fields, and other disturbed sites.
Comments: Naturalized from Eurasia, pusley is one of the most palatable of wild potherbs. It was used as food in India for over two thousand years and by Europeans for hundreds. The fatty or slimy quality is sometimes objectionable, but this problem is overcome by baking the cooked tips with breadcrumbs and a beaten egg. Pusley can also be pickled or used in salads. American Indians used the seeds for making mush or bread.

441 Pokeweed

Phytolacca americana L.
Phytolaccaceae (Pokeweed Family)
Flowers: May–Frost
Description: Robust, perennial herb, 3–10' tall, from a thick root; leaves alternate, entire, smooth; plant unpleasantly scented; berries in long racemes, dark purple when ripe in the fall.
Range-Habitat: Common throughout the state; in the coastal plain in fields, pastures, vacant lots, railroad embankments, abandoned house sites, newly created openings in forests and pastures, and barnyards.

Comments: All parts or the plant are poisonous except the young shoots which are eaten as a potherb before the pink color appears. The juice of the berry was used as a dye by American Indians and early colonists.

442 **Bitterweed**

Helenium amarum (Raf.) H. Rock
Synonym: *Helenium tenuifolium* Nuttall
Asteraceae (Aster or Sunflower Family)
Flowers: May–Frost
Description: Aromatic annual herb to 40" tall with taproot; freely branched above; leaves linear, often with smaller, axillary clusters of leaves; basal leaves soon deciduous.
Range-Habitat: Throughout the state; in the coastal plain common in a variety of disturbed areas and sites with poor soils including pastures, roadsides, abandoned lots, fields, and open woods.
Comments: Bitterweed can be a serious weed in pastures where cattle graze. When other forage is scarce, cattle will graze on bitterweed which gives a bitter taste to their milk. Bitterweed is suspected of being poisonous to humans.

443 **Wood Sage; Germander**

Teucrium canadense L.
Lamiaceae (Mint Family)
Flowers: June–August
Description: Erect, rhizomatous perennial, 1–3' tall; stems four-angled; leaves opposite, lanceolate, toothed, with under surfaces densely hairy.
Range-Habitat: Generally throughout the state; in the coastal plain in a variety of disturbed habitats, moist places in thin woods, marshes, ponds, and meadows.
Taxonomy: Wood sage is a highly variable species. Plants in the coastal area tend to be more hairy that those in the piedmont and mountains.

444 **Woolly Mullein**

Verbascum thapsus L.
Scrophulariaceae (Snapdragon Family)
Flowers: June–September
Description: Densely wooly biennial, 2–6' tall; basal cluster of thick, velvety leaves present only in first year; stem develops the second year with leaves gradually reduced upward; flowers fragrant, appearing the second year.
Range-Habitat: Commonly naturalized throughout the state; in the coastal plain along roads, pastures, fallow fields, sandy open woods, vacant lots, and shell middens.
Comments: Wooly mullein is a native of Eurasia and naturalized in North America. Few plants have had as many former uses as wooly mullein, both in the Old and New World. Roman soldiers dipped the stalks into grease for torches. Indians, plantation slaves, and colonists lined their shoes

with the velvety leaves as cushions and to keep out the cold. Quaker women, not allowed to use cosmetics, rubbed their cheeks with the leaves to achieve a reddish look. A tea from the plant was used to treat colds. American Indians smoked the dried leaves, flowers, or roots for pulmonary ailments, and slaves on southern plantations used a decoction of the plant to relieve the pain of hemorrhoids or to induce sleep. Mullein does have a narcotic quality.

445 **Rattle-bush**

Daubentonia punicea (Cav.) DC.
Synonym: *Sesbania punicea* (Cav.) Benth.
Fabaceae (Pea or Bean Family)
Flowers: June–September
Description: Shrub 3–4' tall; leaves even-pinnately compound with twelve to forty entire leaflets; legume four-winged, matures August–November.
Range-Habitat: Naturalized in the coastal plain; frequent along roadsides and ditches.
Comments: Rattle-bush was introduced into Florida as an ornamental. From there it escaped and spread throughout the Southeast. Kingsbury (1964) reports rattle-bush is highly poisonous to fowl and sheep (and I feel this must also be assumed to be poisonous to humans).

446 **Partridge Pea**

Cassia fasciculata Michaux
Synonym: *Chamaecrista fasciculata* (Michaux) Greene
Fabaceae (Pea or Bean Family)
Flowers: June–September
Description: Annual herb, reclining or erect, to 3' tall; leaves sensitive, folding when touched, even-pinnately compound with twelve to thirty-six leaflets; sessile, depressed, saucer-shaped gland near the middle of the leaf stalk; legume mature July–November.
Range-Habitat: Common throughout the state in various disturbed sites; especially common in abandoned agricultural fields.
Comments: The seeds are a valuable native food for quail.

447 **Sida**

Sida rhombifolia L.
Malvaceae (Mallow Family)
Flowers: June–October
Description: Annual or biennial herb with tough stem to 4' tall; leaves alternate; flowers solitary, axillary, on long stalks; the annual and biennial plants flower in June–October, the biennials again in April–May.
Range-Habitat: Naturalized throughout the piedmont and coastal plain; common in the coastal plain along roads, in fallow fields, yards, pastures, and vacant lots.

448 **Common Evening-primrose**
Oenothera biennis L.
Onagraceae (Evening-primrose Family)
Flowers: June–October
Description: Erect, biennial herb, 2–8' tall, from a taproot; branching only near the top; leaves alternate; flowers in terminal clusters, lemon-scented; capsule cylindrical.
Range-Habitat: Common throughout the state; in the coastal plain found along roadsides; in fallow, sandy fields; in vacant lots; around abandoned buildings; and on stable dunes.
Comments: The common name refers to the flowers which open in the evening and close by noon the next day. The taproot is edible; however, only the newly grown roots can be used and these must be collected in the early spring. The plant has many medicinal uses. For example, Native Americans used a tea from the roots for obesity and bowel pains.

449 **Common Nightshade**
Solanum americanum Miller
Solanaceae (Nightshade Family)
Flowers: June–Frost
Description: Annual, smooth herb, 1–2.5' tall; leaves alternate; fruit a berry, black when ripe; ripe and unripe fruits (green) often occur on the plant at the same time.
Range-Habitat: Common throughout the state; in the coastal plain common in fields, abandoned lots, yards, woodland borders, roadsides, and cultivated areas.
Comments: This species of *Solanum* is native, but as open, disturbed areas were created, it quickly invaded them and became weedy. The leaves and unripe berries contain an alkaloid that has poisoned cattle and humans. Although some sources claim the toxic quality lessens as the berries ripen, it still is best to avoid them.

450 **Mexican-clover**
Richardia scabra L.
Rubiaceae (Madder Family)
Flowers: June–Frost
Description: Weedy annual with diffuse, creeping stems; leaves opposite; outer parts of leaves and lower midrib rough to the touch.
Range-Habitat: Chiefly coastal plain; common in longleaf pine savannas, fields, roadsides, lawns, vacant lots, and roadsides.
Comments: Naturalized from tropical America.

451 **Winged Sumac**
Rhus copallina L.
Anacardiaceae (Cashew Family)
Flowers: July–September
Description: Rhizomatous shrub or small tree, 20–25' tall; stems densely short-hairy; leaves deciduous, alternate, pinnately

compound; midrib of leaf winged; leaves turn bright red in fall; male and female flowers in separate clusters on different plants; drupes in clusters, dark red when ripe in fall.

Range-Habitat: Common throughout the state; in the coastal plain in oak-hickory and pine–mixed hardwood forests, longleaf pine flatwoods, along fence rows and roadsides, and in thickets, pastures, and old fields.

Comments: The hairs on the surface of the drupes contain malic acid, a pleasant tasting acid. Native Americans, and then European settlers, used the fruit as a source of a cool, summer drink. It is prepared by bruising the fruits in water to free the acid and then straining the water (through cloth) to remove the hairs. Sugar can be added; the resulting drink is similar to pink lemonade.

452 Jimson-weed; Jamestown-weed

Datura stramonium L.
Solanaceae (Nightshade Family)
Flowers: July–September

Description: Rank smelling, coarse, annual weed, 1–5' tall; leaves alternate, irregularly lobed; fruit an erect, spiny, many-seeded capsule that matures August–October.

Range-Habitat: Naturalized and common throughout the state; in the coastal plain in barnyards, fields, roadsides, vacant lots, and other disturbed areas.

Comments: Jimson-weed is a corruption of Jamestown-weed, the latter name derived from its growing around Jamestown, Virginia, and a report that soldiers sent there in 1676 to quell the Bacon Rebellion became intoxicated from smoking the plant. All parts of the plant are poisonous; cattle and sheep have been killed by grazing on it, and children have been poisoned by eating the fruit.

There seems to be a question in the literature about where jimson-weed is native. One source says it comes from tropical America; another source says Europe. Another source questions both of these by pointing out that it was reported growing around Jamestown in 1676, suggesting it is native to North America.

453 Rattlebox

Crotalaria spectabilis Roth
Fabaceae (Pea or Bean Family)
Flowers: July–September

Description: Annual herb, 2–3' tall, from a woody taproot; leaves simple, alternate, markedly obovate; stems commonly dark purple; legume inflated, many seeded, becoming black at maturity in August–October.

Range-Habitat: Naturalized and common in the piedmont and coastal plain; in the coastal plain in cultivated and fallow fields, roadsides, and other disturbed areas.

Comments: A native of India, it was introduced as a soil-building green manure. It quickly became naturalized and became a serious pest in agricultural crops. It is poisonous and has caused severe losses in fowl, cattle, horses, and swine. It is also poisonous to humans.

The common name comes from the legume; when mature, the seeds break loose inside and rattle.

454 Common Morning-glory

Ipomoea purpurea (L.) Roth
Convolvulaceae (Morning-glory Family)
Flowers: July–September
Description: Twining, annual, herbaceous vine with one main stem from a taproot; leaves simple, heart-shaped; corolla purple, red, bluish, white, or variegated.
Range-Habitat: Common throughout the state; in the coastal plain in fallow and cultivated fields, roadsides, fence rows, and abandoned lots.
Comments: Common morning-glory is native to tropical America and was introduced into North America as an ornamental. It escaped from gardens and is now widely naturalized.

455 Common Reed

Phragmites australis (Cav.) Trin. ex Steud.
Synonym: *Phragmites communis* (L.) Trin.
Poaceae (Grass Family)
Flowers: July–October
Description: Coarse, rhizomatous perennial, 5–15' tall; often forming large colonies; inflorescence a dense, tawny to purple panicle.
Range-Habitat: Occasional along the coastal area of the state in dredged soil disposal sites, fresh and brackish marshes, pond margins, ditches, sloughs behind coastal dunes, and other disturbed habitats.
Comments: Common reed is found throughout the United States and in South America, Europe, Asia, Africa, and Australia. It began to appear along the South Carolina coast in the 1970s, apparently spread by dredging barges that moved up and down the Intracoastal Waterway carrying rhizomes in the mud that adhered to the barges.

Common reed is a serious weed in wetland areas managed for wildlife because it can dominant the site at the exclusion of wildlife food plants. Common reed itself is not a valuable food source.

Common reed grows so fast and on such a variety of soils that it is being explored as a potential energy crop. Some of the uses worldwide are thatched homes, pulp for paper, making alcohol, biological filter for waste water, cemented reed blocks, synthetic fibers, and insulation material.

456 **Mexican-tea; Worm-seed**

Chenopodium ambrosioides L.
Chenopodiaceae (Goosefoot Family)
Flowers: July–Frost
Description: Typically an annual plant, but in the South it may live for several years; erect herb, 2–4' tall, covered with glandular-resin dots; strongly pungent; leaves alternate; flowers in dense clusters on short to long spikes.
Range-Habitat: Common throughout the state; in the coastal plain in cultivated and fallow fields, pastures, vacant lots, fence rows, railroad embankments, and roadsides.
Comments: Worm-seed is naturalized from tropical America. In earlier times in the Lowcountry it was used as a vermifuge by rural people. Today it is the source of chenopodium oil used to treat domestic stock for intestinal worms and to treat hookworm infections in humans.

457 **Ivy-leaf Morning-glory**

Ipomoea hederacea (L.) Jacquin
Convolvulaceae (Morning-glory Family)
Flowers: July–Frost
Description: Hairy, annual, twining or climbing, herbaceous vine; flower stalks with strongly, backward-bent hairs; corolla tube white, limb at first blue with white center, changing to rose-purple; sepals prominently hairy at the base, with long tapering tips that spread or curve backward.
Range-Habitat: Common weed throughout the piedmont and coastal plain, less so in the mountains; occurs in a variety of disturbed habitats.
Comments: This species was introduced from tropical America and can be a troublesome weed. The common name refers to the similarity of its leaves with English ivy.
Taxonomy: Considerable confusion exists on the taxonomy of this species. Radford et al. (1968) state that there are two varieties: var. *hederacea* with three-lobed leaves (plate 457) and var. *integriuscula* with entire leaves. Godfrey and Wooten (1981) indicate that both types of leaves may occur on the same plant and do not list varieties.

458 **Mistflower; Ageratum**

Eupatorium coelestinum L.
Synonym: *Conoclinium coelestinum* (L.) DC.
Asteraceae (Aster or Sunflower Family)
Flowers: Late July–November
Description: Perennial herb with erect, solid stems, 12–40" tall; leaves opposite, resinous-glandular beneath; densely hairy; forming colonies from rhizomes.
Range-Habitat: Common throughout the coastal plain; occupying a diverse series of natural habitats and disturbed sites, especially roadsides; in general, more common in moist to wet places.

Comments: The common name, mistflower, comes from the combined effect of the conspicuously exerted stigmas. Mistflower is a weedy native and is used commonly as an ornamental.

459 **Jacquemontia; Tie Vine**
Jacquemontia tamnifolia (L.) Grisebach
Convolvulaceae (Morning-glory Family)
Flowers: August–September
Description: Herbaceous, annual, twining vine; tawny hairy, especially the inflorescence; corolla blue, funnel-shaped.
Range-Habitat: Primarily a coastal plain species; occasional in cultivated fields, gardens, and roadsides.
Comments: Tie vine gets its common name because it is a weed in row crops which ties up the crop with its vines and makes it difficult to harvest.

460 **Coastal Morning-glory**
Ipomoea trichocarpa Ell.
Convolvulaceae (Morning-glory Family)
Flowers: August–October
Description: Twining, perennial vine from a branched root; corolla pink to purple, rarely white; leaves ovate, entire, or more often with two basal lobes; sepals ciliate on margin and hairy at base.
Range-Habitat: Common in the outer coastal plain in abandoned fields, roadsides, vacant lots, fence rows, and thickets.

461 **Rabbit Tobacco; Life Everlasting**
Gnaphalium obtusifolium L.
Asteraceae (Aster or Sunflower Family)
Flowers: August–October
Description: Erect, fragrant, winter or summer annual herb; stems erect, 1–3' tall, densely whitish woolly; dead stems tend to remain throughout the winter; stem leaves alternate, green above, whitish woolly beneath.
Range-Habitat: A common native throughout the state; in the coastal plain in fallow fields, pastures, and open, sandy woodlands.
Comments: Rabbit tobacco was once the most popular native cold remedy in coastal South Carolina (Morton, 1974). It is still sold by the "flower ladies" in the marketplace in Charleston. The tea is drunk as a febrifuge. The tea is bitter and lemon juice is added to make it palatable for children. The plant is also smoked for asthma by rural people today.

Historically, Francis Porcher (1869) said the plant was an astringent, and the leaves and flowers were chewed and the juice swallowed to relieve ulcerations in the mouth and throat.

462 **Boneset; Thoroughwort**
Eupatorium perfoliatum L.
Asteraceae (Aster or Sunflower Family)
Flowers: August–October
Description: Perennial herb with erect, solid stems, 2–4' tall; often growing in clumps; leaves opposite, perfoliate; plant conspicuously hairy.
Range-Habitat: Throughout the state; in the coastal plain common in damp, moist areas such as wet meadows, alluvial woods, marshes, ditches, and pastures.
Comments: Boneset has been used in a number of folk remedies in the Lowcountry and elsewhere. The common name comes from the belief that the plant could cause rapid union of broken bones as suggested by the basal union of each pair of opposite leaves. Francis Porcher (1869) reported it was employed by the slaves on plantations as a diaphoretic in colds, fevers, and typhoid pneumonia. Also, a tea from the leaves was a substitute for quinine to treat malaria. In Appalachia, a tea from the leaves is used as a laxative and as a treatment for coughs and chest illnesses.

463 **Chocolate-weed**
Melochia corchorifolia L.
Sterculiaceae (Cacao Family)
Flowers: August–October
Description: Annual herb with tap roots; stems tough, with remote, star-shaped hairs; leaves simple, alternate; flowers mostly in dense, terminal heads.
Range-Habitat: Chiefly a coastal plain species; weedy and occasional in sandy fields.

464 **Gerardia**
Agalinis purpurea (L.) Pennell
Scrophulariaceae (Snapdragon or Figwort Family)
Flowers: August–Frost
Description: Annual with abundantly branched, spreading, upper stems, 1–4' tall; stem smooth or weakly rough; leaves opposite, linear; corolla throat lined with yellow.
Range-Habitat: Common throughout the state in moist to wet sites in a variety of natural and ruderal gardens such as longleaf pine savannas and flatwoods, bogs, pond margins, interdune swales, roadsides, and forest margins.
Taxonomy: *A. purpurea* is similar to *A. fasciculata;* it is difficult to distinguish between the two in the field. They grow in similar habitats and have essentially the same flowering times. The latter has primary leaves with well-developed axillary fascicles and rough stems while the former is without obvious axillary fascicles and has smooth or weakly rough stems.
Comments: Gerardia is a native species and the most abundant species of *Agalinis* in the state. It has become weedy and has

invaded a variety of disturbed sites, being especially common along roadsides. Like all species of *Agalinis,* it is believed to be a hemiparasite on roots of grasses or other herbs.

465 **Red Morning-glory**

Ipomoea coccinea L.

Convolvulaceae (Morning-glory Family)

Flowers: August–Frost

Description: Annual, herbaceous, twining vine; leaves ovate with acuminate tip, basal lobes rounded or with angular projections; corolla scarlet with a long tube.

Range-Habitat: Chiefly a piedmont and coastal plain species; common in the coastal plain in fallow and cultivated fields, roadsides, thickets, and abandoned lots.

Comments: Native to tropical America, red morning-glory was introduced into gardens where it escaped and became widely naturalized.

466 **Cypress-vine**

Ipomoea quamoclit L.

Convolvulaceae (Morning-glory Family)

Flowers: September–Frost

Description: Twining, herbaceous, annual vine; leaves pinnately divided into narrow, linear segments; corolla crimson with a long tube.

Range-Habitat: Chiefly coastal plain; occasional along roadsides, fences, and cultivated fields.

Comments: Cypress-vine is naturalized from tropical America. The common name refers to leaves that resemble the terminal, needle-bearing twigs of cypress trees.

PART THREE

Guide to Sites

Guide to Selected Sites of Wildflower Gardens

This section is a field guide to forty-eight sites that harbor wildflower gardens readily available to Lowcountry residents. The sites are arranged alphabetically by county. The numbers following each site correspond to the locations on the map of the Lowcountry and Lower Pee Dee. The table can be used to quickly find which gardens occur at a particular site, or to find which sites harbor a particular natural garden. Throughout the field guide, common names are used for species pictured in this book; for other species the scientific name follows the common name in parenthesis.

The following information is included for each site: (a) directions to the site (or information on where directions may be obtained); (b) a list of its wildflower gardens; (c) any unusual aspects of the site (for example, rare plants or gardens) based on my observations; (d) a representative list of species; and (e) information about individuals, agencies, or foundations to contact to obtain access to sites (other than state parks or heritage preserves).

State parks and heritage preserves represent significant wildflower sites in the Lowcountry because they harbor a diversity of natural gardens, and they are readily available year around. For information on the state parks, contact:

S.C. State Parks System
1205 Pendleton Street
Columbia, SC 29202
Telephone: 803-758-3622

For information on heritage preserves, contact:

S.C. Department of Natural Resources
Wildlife Diversity Section
Heritage Trust Program
Post Office Box 167
Columbia, SC 29202
Telephone: 803-734-3918

The descriptions of sites that follow are based on observations from 1988 to 1993. As each year passes, changes occur, from both natural and human-made sources. The end of a barrier island that has accreted and harbors a coastal dune community one year may undergo erosion the next year, thereby destroying the dune system. Or pine savannas that harbor typical showy herbs one year may go unburned for several years, thus eliminating or reducing the showy species. Even yearly changes occur with some species although the community may not change. A longleaf pine savanna may harbor a large population of snowy orchid one year; the next few years none may be present.

Private properties such as hunting preserves or plantations cannot be listed; this would be a violation of private trust. However, individuals who have access to large private holdings should readily make use of these sites. Most often these areas have prescribed burning programs that favor wildflower gardens (for example, pine savannas maintained for hunting) or old-growth forests that harbor a diversity of wildflowers.

Selected highways, although they pass through private property, are listed because they traverse rich communities; their banks, which are in the public domain, harbor species of the adjacent gardens.

The river gardens present a different situation. Navigable waters are in the public domain; often, however, the adjacent marshland is privately owned and posted. One may view the marsh wildflowers from a boat, but one must use discretion on entering the marsh by foot. It is beyond the limitations of this book to include a map of all the river systems showing which lands are private or public.

BAMBERG COUNTY

Cathedral Bay Heritage Preserve: Site 1

Although Cathedral Bay is located in the inner coastal plain, a few miles above the boundary of the outer coastal plain, it is included here because it is a striking example of a pond cypress savanna within a clay-based Carolina bay. Cathedral Bay is also known as Chitty Bay or Chitty Pond. To reach Cathedral Bay, turn east off U.S. Highway 321 at Olar onto S.C. Highway 64, and go 1.67 miles. The bay is adjacent to the highway; only the bay, with clearly marked boundaries, is public; the adjoining land is private.

The bay harbors a pond cypress savanna. During dry seasons, one may easily walk into the interior; during wet seasons, one may canoe through the bay. Canoes can easily be put in from Highway 64. No motorized boats are allowed.

Pond cypress dominates the canopy; also present are swamp-gum and sweet-gum. This savanna is not rich in wildflowers, but one can see tall milkweed and Virginia meadow-beauty. The sedge that dominates the herbaceous layer is Walter's sedge (*Carex walteriana*).

Rivers Bridge State Park: Site 2

Rivers Bridge State Park lies adjacent to the Salkehatchie River—a coastal, blackwater river-swamp that has its origin in Barnwell County south of Williston. The 390-acre state park is the only state park that commemorates the Civil War. The breastworks that guarded the river crossing from Sherman's advancing Union troops today overlook the haunting, beautiful river swamp the Union troops had to forge. At the end of the road that leads to the breastworks, one can see the earthen causeway that was built through the swamp long ago for people to cross the Salkehatchie River. The name of the park, Rivers Bridge, comes from the sixteen bridges built along the causeway to cross water-flows.

The Salkehatchie River floodplain adjacent to the breastworks supports an alluvial swamp forest with large specimens of bald cypress. Other species of note include red maple, climbing hydrangea, sweet bay, cross vine, Virginia willow, American holly, Spanish moss, red bay (*Persea borbonia*), butterweed, American elm (*Ulmus americana*), ironwood (*Carpinus caroliniana*), mistletoe, and dog-hobble along the edge of the swamp.

A rare swamp community, a stream-side bay forest, occurs at a slightly higher elevation above the river swamp at the end of the breastworks. Dog-hobble forms a dense thicket as almost the only shrub. The canopy is made of large specimens of loblolly bay, red bay, tulip tree, bull bay, and American holly.

BEAUFORT COUNTY

Green's Shell Enclosure Heritage Preserve: Site 3

Green's Shell Enclosure represents an Indian midden originally made in a maritime forest on the edge of a saltwater creek. It is located on Jenkins Island (adjacent to Hilton Head Island) and is reached by turning off U.S. Highway 278 onto Squire Pope Road. Travel down this road until you come to the sign for Davis Landscape, Inc. The unpaved road leading to the midden is just before the Davis sign and ends at a green pipe gate. It is a short walk, past two cemeteries, to the midden.

Five natural gardens can be seen at this preserve. A *Spartina* salt marsh flanks the entire site. Several salt flats occur in the marsh with perennial glasswort. A salt shrub thicket with sea ox-eye, seaside goldenrod, marsh-elder, and groundsel-tree lies above the marsh line.

Most of the brush on the midden and maritime forest is kept cut, but numerous species of the maritime forest can be seen. Look for live oak, yaupon, Spanish moss, cabbage palmetto, Carolina laurel cherry, trailing bluet, bull bay, coral bean, Spanish bayonet, tough bumelia (*Bumelia tenax*), southern red cedar, and cross vine. On the shell mounds look for a rare tree, buckthorn, and hackberry (*Celtis laevigata*).

Over the years, numerous weedy species have invaded the midden and surrounding area: privets, Japanese honeysuckle, and wooly mullein are examples.

Hunting Island State Park: Site 4

The southern and northern ends of Hunting Island are accreting and offer the typical coastal beach and dune gardens. The northern dune system is unusually high for the Carolina coast; some dunes are more than 20 feet high. Salt flats, salt marshes, and salt shrub thickets occur at each end and along the backside of the island.

The maritime forest was burned severely in the 1940s, and large trees are not common. The forest today is dominated by slash pine (*Pinus elliottii*) and cabbage palmetto with saw palmetto common in the shrub layer. The dominance of even-age pine on Hunting Island is a result of the fire; likewise, the absence of large live oaks is probably due to the fire. A well-marked nature trail passes through the maritime forest. Relict dunes that form the barrier islands are visible along the length of the trail.

Victoria Bluff Heritage Preserve: Site 5

The entrance to the road (State Highway 744) leading through Victoria Bluff Preserve is off U.S. Highway 278, just west of Hilton Head Island. Signs clearly mark the entrance. A map of the Preserve is probably necessary for visitors to locate the many trails and to define the boundaries.

Victoria Bluff Preserve consists of approximately 1,000 acres and contains a mix of natural gardens. The heart of the Preserve is the pine–saw palmetto flatwoods which are rare in South Carolina. Longleaf pine and slash pine (*P. elliottii*) are the dominant trees. Controlled burning has been used to maintain this fire dependent community. Other species to look for are black-root, bracken fern (*Pteridium aquilinum*), stagger-bush, yellow jessamine, hairy wicky, horse sugar, silver-leaved grass, inkberry (*Ilex glabra*), species of aster, and vanilla-plant (*Trilisa odoratissima*). The most open stands of the flatwoods occur along U.S. 278 and along Highway 744 beginning 0.4 miles from U.S. 278.

Other communities to look for are pond pine woodlands with associated pocosin species including fetterbush, red bay, inkberry, wax myrtle, saw palmetto, and the rare *Lyonia ferruginea;* bay forests with pond pine and loblolly in the canopy and the shrubs red bay, wax myrtle, fetterbush, and *Lyonia ferruginea;* inland maritime forests with live oak and cabbage palmetto; upland swamp forests with red maple and sweet-gum; an upland pond with scattered button-bush, bog-mat, duckweed, and frog's-bit; swamp-gum savannas with the rare pond spice (*Litsea aestivalis*), red bay, Virginia chain-fern (*Woodwardia virginica*), and fetterbush on the edge; and a laurel oak forest with red bay, loblolly pine, sparkleberry, bull bay, saw palmetto, water oak (*Quercus nigra*), and wax myrtle.

BERKELEY COUNTY

Bluff Plantation Wildlife Sanctuary: Site 6

The Bluff Plantation Wildlife Sanctuary is located south of Moncks Corner on the Western Branch of the Cooper River. It is reached by taking State Secondary Road 9 off U.S. Highway 52; then taking Secondary Road 260 off Highway 9, which dead ends at the Bluff entrance.

The Sanctuary is owned by the Kathleen O'Brien Foundation of New Orleans, Louisiana. It was established by Miss O'Brien who bequeathed the property to the Foundation. The public is welcome to visit the Sanctuary year around; however, the number of visits per day is limited due to such factors as bad roads in rainy weather or prescribed burning. Call the Sanctuary manager at 803-761-2020 for information about visitation.

Hurricane Hugo severely impacted the upland forests of the Bluff; even today (1994), ingress for wildflower viewing is difficult. However, numerous sites offer excellent wildflower viewing on this 2,000-acre Lowcountry plantation.

The Reserve Swamp represents an old reservoir for the inland rice fields. It harbors an alluvial swamp forest with tupelo gum, bald cypress, red maple, swamp-gum, American elm (*Ulmus americana*), ash, and swamp cottonwood (*Populus heterophylla*). Numerous flowering herbs of the floating aquatic and freshwater marsh gardens occur here. Examples include mosquito fern (*Azolla caroliniana*), all four genera of Lemnaceae, floating bladderwort, frog's-bit, blue flag iris, lizard's tail, and false nettle. The swamp offers an excellent "swamp stomp"; even when flooded, the water level is only waist high.

The abandoned rice fields harbor tidal freshwater marshes. Look for native wisteria, indigo-bush, alligator-weed, arrow arum, swamp rose, obedient plant, water hemlock, pickerel-weed, wapato, ground-nut, water parsnip, swamp rose mallow, hairy hydrolea, cardinal flower, fragrant ladies' tresses, climbing aster, bur-marigold, and water-spider orchid.

At the longleaf pine savanna (pitcher-plant bog on the map of the Bluff), one may find hooded pitcher-plant, trumpet pitcher-plant, sun-bonnets, sundew, colicroot, blue-hearts, orange milkwort, milkweed, smooth meadow-beauty, false asphodel, toothache grass, pine lily, eryngo, Barbara's-buttons, and crested fringed-orchid.

Cooper River: Site 7

The Cooper River and its tributaries (Huger and Quinby creeks which form the Eastern Branch, and Wadboo Creek and the Tailrace Canal which form the Western Branch) harbor hauntingly beautiful wildflower gardens. The abandoned rice fields harbor tidal freshwater marshes. Throughout the growing season a progression of wildflowers can be found. To name a few: pickerelweed, obedient plant, and spider-lily in May; swamp rose mallow and seashore mallow in June and July; cardinal flower in July and August; and climbing aster in September and October. At high tide one can easily enter the abandoned rice fields through breaks in the banks and float in the quiet waters of the fields. The abandoned banks also harbor numerous herbaceous and woody wildflowers.

The typical zonation occurs as one goes down the river: the tidal marsh grades into the brackish marsh which grades into the salt marsh near Charleston Harbor.

There are numerous boat landings along the Cooper River and its tributaries that allow public access; one may refer to a county road map for this information.

Francis Marion National Forest: Site 8

The Francis Marion National Forest consists of approximately 250,000 acres in Berkeley and Charleston counties. Because of its size, most of the natural gardens listed in this book occur in the Francis Marion National Forest. Visitors may obtain a map of the forest by writing or visiting one of two Ranger Stations:

Wambaw Ranger District	Witherbee Ranger District
P.O. Box 106	HC 69, Box 1532
McClellanville, SC 29458	Moncks Corner, SC 29461

The Witherbee District covers only Berkeley County, while the Wambaw District includes lands in both Berkeley and Charleston counties.

For each natural garden below, only a few sites harboring the most accessible and best examples of each are given. It would be impractical to list every site in the Forest for each natural garden.

Beech Forests

Upland, deciduous forests such as the beech and oak-hickory forests suffered the most damage from Hurricane Hugo in 1989. The canopies of these forests were virtually removed, and the invasion of briers and weedy species makes ingress difficult. One exceptional beech forest occurred along the southern side of Huger Creek on the western side of S-402. Over a hundred species of wildflowers had been identified in this beech forest before the hurricane. With modest effort, one may fight through the vines and briers and find pockets of beech trees still standing. Numerous spring ephemerals, which are adapted to sunny conditions before the canopy develops, have persisted since the hurricane and include may-apple, little sweet betsy, green-and-gold, violet wood sorrel, wild geranium, and bloodroot. No other stands of beech forests in the Forest are suitable for field trips at the present time.

Bald Cypress–Tupelo Gum Swamp Forests

The Santee River is a major brownwater river of the United States. Much acreage of bottomland bald cypress–tupelo gum swamp forests occurs along its length where it passes through the Francis Marion National Forest. A good site to see this community is at Lake Guilliard Scenic Area and Natural Area east of Jamestown. At either end of the lake, one can follow a series of sloughs that harbor large bald cypress and tupelo gum trees. Cypress knees, some as high as 6 feet, are evident.

Hardwood Bottoms

Significant stands of hardwood bottoms, like the swamp forests above, occur along the Santee River. Hardwood bottoms can also be found at Lake Guilliard Scenic Area and Natural Area at elevations slightly above the swamp forests.

Longleaf Pine Flatwoods

Several roads provide access to mature stands of longleaf flatwoods. Five examples are (1) State Highway 41 for approximately 5 miles north of the Wando River; (2) Forest Route 173; (3) State Highway 171 northwest of Witherbee (at the intersection of S-171 and Forest Route 132, look for American chaffseed in May and June); (4) State Highway 98 for about 5 miles north of Wando; and (5) State Highway 654 (Halfway Creek Road) southwest of Honey Hill.

Pocosins

An outstanding example of a Carolina bay with pocosin vegetation is Big Ocean Bay Scenic Area northeast of Bethera. The site is actually two Carolina bays; the smaller bay, Little Ocean Bay (figure 2), is adjacent to Forest Route 110 and harbors a low pocosin throughout the bay. In its center look for the rare leather-leaf, and honeycup and witch-alder along the edge.

Pond Cypress Savannas

Many examples of pond cypress savannas occur in the Francis Marion National Forest. Directions to two accessible sites are as follows: (1) take State Highway 654 southwest of Honey Hill to its intersection with Forest Route 201-B; at this point on the northwestern side of 654 is a pond cypress savanna with numerous wildflowers including pine lily, sneeze-weed, trumpet pitcher-plant, and awned meadow-beauty; and (2) north of Cainhoy, turn east off State Highway 41 onto Forest Route 183 (Hoover Road), go about a 100 yards, look to the left and you will see the savanna through a stand of pines; species to look for are yellow trumpet pitcher-plant, bay blue-flag iris, tickseed, and pipewort.

Pond Cypress–Swamp-gum Swamp Forests

An example of this community dominated by swamp-gum is located northwest of the intersection of Forest Route 166 and

State Secondary 10-98 (Halfway Creek Road). The swamp is in the corner formed by the power line, Route 166, and S-10-98. The interior of the swamp opens up and wading is easy. Pond spice (*Litsea aestivalis*) is common within the interior, and the endangered pond berry occurs on the southern margin.

A series of lime sinks around Honey Hill harbor swamps dominated by pond cypress. In many of these sinks, one can find pond spice, the endangered pond berry, and a variety of wildflowers. This area is being considered for designation by the Forest Service as a Research Natural Area. Persons interested in visiting these sinks may inquire at the office of the Wambaw Ranger District for directions. The roads leading to the sinks have no sign designations, so it is difficult to give directions in this guide.

Stream Banks

Examples of this community can be seen along the Santee River in the vicinity of Alvin.

Xeric Sandhills

The best example of the xeric sandhills occurs at Cedar Hill, reached by turning off Echaw Road (Forest Route 204) onto Forest Route 204-D. The xeric community occurs to the right. Species of note are tread-softly, sandhills milkweed, thistle, gerardia, and wire plant (*Stipulicida setacea*).

Old Santee Canal State Park: Site 9

This 200-acre state park is located southeast of the junction between the Tailrace Canal and U.S. Highway 17 near Moncks Corner and encompasses part of Stoney Landing Plantation. The site is extremely rich in wildflowers with many diverse natural communities. Although Hurricane Hugo caused severe damage to the upland communities, it is still too early to evaluate the effects on the herbaceous flora.

The most significant natural garden is the marl forest that occurs on the bluff flanking Biggin Creek. The nature trail only passes along a short portion of the bluff and more serious botany students will have to get permission from the park ranger to botanize the remainder of the marl forest. Numerous trees characteristic of marl forests can be found: hop hornbeam, white basswood, buckthorn, yellow chestnut oak (*Quercus muehlenbergii*), the rare nutmeg hickory (*Carya myristicaeformis*), southern sugar maple (*Acer saccharum* subsp. *floridanum*), and slippery elm (*Ulmus rubra*). Look for the following herbs: meadow parsnip, alumroot, thimbleweed, horse-balm, elytraria, and widespread maiden fern (*Thelypteris kunthii*).

The bluff also contains elements of the beech and oak-hickory forests with numerous showy wildflowers and shrubs: wild ginger, may-apple, pawpaw, sweet shrub, bloodroot, red buckeye, fringe-tree, Indian pink, coral honeysuckle, and lopseed.

Along Biggin Creek look for the following tidal marsh and swamp species: swamp rose, pickerelweed, indigo-bush, alligator-weed, water-spider orchid, arrow arum, water hemlock, obedient plant, swamp rose mallow, cardinal flower, climbing aster, bald cypress, red maple, Virginia willow, and supplejack.

Wadboo Creek Marl Forest: Site 10

The marl forests along Wadboo Creek are one of the most significant botanical sites in the coastal area. It is property of the Santee Cooper Public Service Authority, headquartered in Moncks Corner, and is open to the public. Unfortunately, the only access for the public is by boat because the site is bordered by private property. One must put in by boat at the Dennis Landing on State Highway 402 east of Moncks Corner. Approximately a quarter of a mile north of the landing will appear a steep, limestone bluff. Just past the bluff the land slopes to the river and one can land a boat. The marl forest occurs on the bluff that runs along the creek floodplain.

Numerous uncommon and rare trees and herbs occur in this marl forest. Trees include yellow chestnut oak (*Quercus muehlenbergii*), hop hornbeam, white basswood, slippery elm (*Ulmus rubra*), buckthorn, southern sugar maple (*Acer saccharum* subsp. *floridanum*), redbud, flowering dogwood, and red cedar. Rare or uncommon herbs include bloodroot, meadow parsnip, thimbleweed, elytraria, autumn coral-root (*Corallorhiza odontorhiza*), wild ginger, liverleaf (*Hepatica americana*), widespread maiden fern (*Thelypteris kunthii*), and incised groovebur (*Agrimonia incisa*).

CHARLESTON COUNTY

Accreted Beach at Sullivans Island: Site 11

Accreted Beach is a section of beach and dunes owned by the Town of Sullivans Island and open to the public year-round. It is reached from either Station 18 or Station 16; limited parking is available at the end of each road. Trails lead from the end of the roads to the beach.

Sea rocket and Russian thistle are both found in the beach community. The well-developed dunes support many of the typical dune plants: devil-joint, seaside pennywort, beach evening-primrose, gaillardia, silver-leaf croton, sea purslane, dune spurge, beach pea, sea oats, camphorweed, dune sandbur, seaside panicum, seashore-elder, and beach morning-glory. The maritime shrub thickets support typical species such as wax myrtle and dune greenbrier; however, numerous weedy species, such as rattle-bush and giant foxtail (*Setaria magna*), have invaded the thicket after disturbance by Hurricane Hugo.

Audubon Swamp Garden at Magnolia Gardens: Site 12

Audubon Swamp Garden is located at Magnolia Gardens on State Highway 61 north of Charleston. An admission fee is required.

Wildflowers from three natural gardens can be found at the Audubon Swamp Garden: a swamp forest dominated by red maple and swamp-gum, freshwater marsh, and freshwater aquatics. The visitor should understand, however, that the majority of plants along the trails are horticultural species planted to enhance the beauty of the swamp. Freshwater aquatics include species of Lemnaceae, fragrant water-lily, and floating bladderwort. Marsh plants include native wisteria, alligator-weed, common cat-tail, water willow, climbing hempweed, and bur-marigold. Swamp species include red maple, bald cypress, swamp dogwood, Virginia willow, lizard's-tail, tupelo gum, and climbing hydrangea. Additional native plants include sweet bay, button-bush, swamp willow (*Salix caroliniana*), Virginia creeper, Jack-in-the-pulpit, and false nettle on fallen logs and buttresses.

Beachwalker County Park: Site 13

Beachwalker County Park is located on the western end of Kiawah Island and is operated for the public through a cooperative effort of the Kiawah Island Company, Charleston County, and the Charleston County Parks and Recreation Commission. Information about the park can be obtained by calling 803-762-2172. The park is closed from November through March.

The park is located on the southwestern end of the island on an accreting beach ridge which is approximately 1 mile long. On the ocean side are the coastal beaches and coastal dunes that include such species as seaside panicum, dune sandbur, beach evening-primrose, camphorweed, sea oats, seaside pennywort, sea rocket, Russian thistle, seashore-elder, silver-leaf croton, beach morning-glory, dune spurge, beach pea, and frog-fruits. A significant population of sweet grass occurs in the flats between the dunes. Horseweed and common marsh-pink grow in the wet swales behind the dunes.

The stabilized dunes on the ridge support typical maritime shrub thickets dominated by wax myrtle. Creeping cucumber (*Melothria pendula*), an annual vine climbing by tendrils, covers much of the wax myrtle in late summer to fall.

One can walk down the beach to the end of the ridge, then turn back east along the back of the ridge which is bordered by the Kiawah River. Along the river one can see salt marshes, salt flats, and salt shrub thickets. In the fall, sea lavender blooms on the edge of the salt flats and in the higher salt marsh.

Capers Island Heritage Preserve: Site 14

Capers Island is a 2,100-acre barrier island with about 3 miles of front beach. It is one of the chain of islands north of Charleston: Sullivans, Isle of Palms, Dewees, Capers, and Bull.

There is no bridge to Capers Island. The most convenient boat ramp is a public ramp at Breach Inlet on the Isle of Palms; from there it is only a 20-minute run up the Intracoastal Waterway to Capers Island. On the southern end is a large dock and two docking platforms. A well-kept trail, approximately a mile long, leads from the dock to the front beach.

The southern end of this island offers a rich variety of seaside wildflowers. In the vicinity of the dock are salt marshes, salt flats, and salt shrub thickets with smooth cordgrass, sea lavender, perennial glasswort, saltwort, sea ox-eye, seaside goldenrod, marsh-elder, groundsel-tree, and yaupon.

The trail passes three brackish impoundments dominated by needle rush. Blinds along the edge of the impoundments provide camouflage for bird watching. Several naturalized species occur along the trail; look for popcorn tree, Mexican-tea, and pokeweed. Hercules'-club is also common along the trail.

The trail winds along the edge of the maritime forests and through secondary dunes, terminating at the beach where one can see sea rocket and Russian thistle. The dunes offer devil-joint, seaside pennywort, silver-leaf croton, dune spurge, *Yucca gloriosa,* sea oats, camphorweed, beach morning-glory, sea-shore-elder, tough bumelia (*Bumelia tenax*), dune sandbur, trailing bluet, seaside panicum, and sweet grass. Common marsh-pink grows in the dune swales.

The maritime forests suffered severe tree loss from Hurricane Hugo. So much underbrush has developed that it is almost impossible to walk through the forests on the ocean side. The forest on the back side is not as severely disturbed but is still a challenge to enter. Many of the maritime plants can be seen from the trail: live oak, cabbage palmetto, and Spanish moss.

Dill Wildlife Refuge: Site 15

The 600-acre Dill Wildlife Refuge on James Island is owned and operated by the Charleston Museum. The nature center and trail system were not complete at publication. Access to the refuge can be arranged by calling the Charleston Museum (803-722-2996).

The Dill Refuge has had an extensive history of land use. All forested areas are secondary growth with signs of repeated disturbance. The dominant wildflower areas are the ruderal gardens. The abandoned tomato fields, hedge rows, and lawn along the main avenue harbor a rich variety of native and introduced weedy species.

Where the northern part of the Refuge lies adjacent to the salt marsh, one can see the salt flats with perennial glasswort, saltwort, and sea lavender.

Folly Beach County Park: Site 16

Folly Beach County Park is located at the southern end of Folly Island. It is open year around; however, the hours change with the seasons. Visitors may call 803-762-2172 for additional information.

The main wildflower gardens of the park are the dunes which begin at the park entrance; the dunes follow the spit that forms the park, then make a sweeping arch at the southern tip and extend along the Folly River for a short distance. Erosion is evident along most of the dunes. Typical coastal dune wildflowers are present. Four species are especially prominent: beach morning-glory, beach pea, beach evening-primrose, and toothache-tree.

Other coastal gardens on the park include salt flats, salt marshes, maritime shrub thickets, and salt shrub thickets.

Francis Marion National Forest: Site 8

Portions of the Francis Marion National Forest in Charleston County lie along the coast and harbor examples of the seaside gardens. They are not, however, readily accessible. Since they are covered in numerous sites in this field guide, these portions are omitted here. One exception is the Sewee Indian Midden.

Longleaf Pine Flatwoods

Stands of longleaf flatwoods are common in the Wambaw District. A newly acquired tract, South Tibwan on U.S. Highway 17 south of McClellanville, harbors a mature, flatwoods community. Also, Forest Route 211 north of State Road 45 passes through stands of longleaf pine flatwoods.

Pocosins

Situated on both sides of Secondary Road 1032 southeast of the county line is a vast pocosin. Species of note are trumpet pitcher-plant, loblolly bay, sweet bay, pond pine, honeycup, rose pogonia, highbush blueberry, and the rare white arum.

Pond Cypress–Swamp-gum Swamp Forests

East of McClellanville, in the angle formed by the intersection of U.S. Highway 17 and State Road 857, is an upland swamp dominated by swamp-gum with pond cypress present. Here one will find the rare, climbing fetterbush growing under the bark of pond cypress.

Shell Mounds

Sewee Indian Midden can be reached by taking Forest Route 243, then find the marked trail to the right leading to the midden. This Indian midden is a protected archeological site. The rare shell-mound buckthorn grows on the mound. Also look for crested coral-root orchid, white basswood, and buckthorn.

Tidal Freshwater Marshes

The best examples of tidal freshwater marshes occur along Wambaw Creek from Still Landing (at the end of Forest Route

211-B) to the boat ramp on Forest Route 204. Wambaw Creek and the adjoining swamp is a nationally designated Wilderness Area (Wambaw Creek Wilderness Area). Look for green-fly orchid on cypress branches and spider-lily on the creek bank.

Tidal Freshwater Swamp Forests

The tidal swamp forests along Wambaw Creek from Still Landing to Forest Route 204 are secondary forests growing on abandoned rice fields. The banks and trunks of the rice fields can still be seen. Swamp-gum is the dominant tree; other species include bald cypress, red maple, water ash, cottonwood, and red bay. Button-bush, swamp rose and Virginia willow are common shrubs. The herbaceous flora is a mix of tidal marsh and swamp species. Look for lizard's tail, golden-club, butterweed, water willow, purple lobelia, coral greenbrier, pickerelweed, water hemlock, arrow arum, and southern rein-orchid (*Habenaria flava*).

Hampton Plantation State Park: Site 17

A plethora of native wildflowers occurs in the natural gardens of this plantation which was home to the former poet laureate of South Carolina, Archibald Rutledge. It is easy to see why he and his family ensured its survival in its natural state by selling it to the state.

At the time of publication of this book, the trail system was not fully established. Visitors must make an appointment with the park manager to have access to most of the wildflower gardens listed below. Call 803-546-9361.

Northeast of the entrance are extensive longleaf pine flatwoods with a rich variety of wildflowers: stagger-bush, black-root, goat's rue, stiff-leaved aster, sweet pepperbush, lead plant, pencil flower, cinnamon fern, and deer's-tongue. Laced throughout the flatwoods are low areas supporting longleaf pine savannas. Look for orange milkwort, colicroot, smooth meadow-beauty, trumpet pitcher-plant, false asphodel, dwarf milkwort, and crested fringed-orchid. Scattered throughout the flatwoods and savannas are pocosins with pond pine, fetterbush, sweet bay, bamboo-vine, and possum-haw.

Near the plantation house is the Small Rice Field which harbors a freshwater marsh with pickerelweed, water hemlock, common cat-tail, marsh bulrush, wild rice, and swamp rose mallow. South of the house is the cypress pond which is actually an upland swamp forest dominated by swamp-gum (*Nyssa biflora*) and pond cypress. Species of note include Virginia willow, sweet bay, styrax, dahoon holly, and resurrection fern. East of the house is the Old Rice Field, a large area presently impounded and covered with a variety of marsh species: pickerelweed, alligator-weed, lizard's tail, button-bush, and southern wild rice. The dike harbors a mixture of trees and shrubs. Look for Cherokee rose and coral honeysuckle.

Surrounding the house are lawns and disturbed areas with ornamental plants and a wide variety of naturalized and native weedy species.

Santee Coastal Reserve: Site 18

The Santee Coastal Reserve (SCR) is owned and operated by the S.C. Department of Natural Resources and consists of approximately 24,000 acres. Contained within the Reserve is Washo Reserve, a 1,040-acre natural area owned by the Nature Conservancy of South Carolina. The SCR is located on the South Santee River and is reached by Secondary Road 857 off U.S. Highway 17 about 52 miles north of Charleston. Information about the Reserve may be obtained from:

Santee Coastal Reserve Manager
P.O. Box 107
McClellanville, SC 29458
803-546-8665

The dominant upland natural gardens of the Reserve are the longleaf pine flatwoods maintained by prescribed burning. The main road leading into the Reserve passes through the flatwoods which harbor typical flatwoods species: blazing-star, black-root, sweet pepperbush, lead plant, pencil flower, deer's-tongue, and cinnamon fern. Laced throughout the flatwoods are longleaf pine savannas with a rich display of herbs: hooded pitcher-plant, yellow meadow-beauty, orange milkwort, colicroot, white sabatia, false asphodel, toothache grass, redroot, trumpet pitcher-plant, and crested fringed-orchid.

Three Carolina bays occur on the Reserve; two harbor upland pond cypress–swamp-gum forests and one a pond cypress savanna.

Washo Reserve is a former rice reserve created by damming a creek. It harbors three natural gardens: an alluvial bald cypress–tupelo gum swamp forest, the freshwater aquatics, and a freshwater marsh. An 800-foot boardwalk traverses these communities. One can see species of Lemnaceae, floating bladderwort, and fragrant water-lily representing the aquatics. Swamp species include red maple, tupelo gum, bald cypress, Virginia willow, supplejack, blue flag iris, and coral greenbrier. Marsh species include alligator-weed, arrow arum, water hemlock, pickerelweed, and water willow.

The Washo Reserve Nature Trail takes the visitor through a transitional maritime forest that includes live oak, Spanish moss, resurrection fern, beauty-berry, American holly (*Ilex opaca*), and bull bay. Transitional species include sweet-gum, oaks, and sparkleberry. The trail then passes around an impoundment harboring a brackish marsh with common cat-tail and soft-stem bulrush.

Included in the Reserve are two barrier islands, Murphy and Cedar, which must be reached by boat. All the maritime communities are represented on these islands.

Tea Farm County Park: Site 19

The development of the county park at Tea Farm on U.S. Highway 17 in Ravenel began in 1993; it will not be completed by the time this book goes to press. The name used above may not be the official name chosen for the park, but it contains enough elements of the name to be used so that readers will have no problem finding the park. Information may be obtained by calling 803-762-2172.

The natural and cultural aspects of this tract of land are exceptional; this park may well become a focal point of the Lowcountry. Disturbance will be held to a minimum, and the natural and cultural aspects will be protected and featured through nature trails and a visitor's center.

A floristic inventory conducted before the park was developed identified 353 different vascular plants, including fifteen species of ferns and seven species of orchids. One hundred thirty-seven of the 353 species are pictured in this book—a main reason the park is included in this field guide. A mix of natural and disturbed, successional communities comprises the park. Walking trails will traverse many of these communities.

A major feature of the park is the inland rice system. Workers have mapped out the complex system of canals, fields, reservoirs, crossbanks, flanking banks, and water control structures that formed the system. One trail will conduct visitors through the rice fields and reservoirs. It is within the two reservoirs that mature, alluvial bald cypress–tupelo gum swamp forests occur. The West Reservoir is significant with large trees present. The former rice fields are in various stages of succession and harbor a mixture of marsh and swamp species. Of note are cardinal flower, purple bladderwort, frog's-bit, blue flag iris, bog-mat, duckmeat, bur-weed, water willow, and fragrant water-lily.

Small stands of beech forests occur scattered throughout the uplands with species such as heart-leaf, cancer-root, spotted wintergreen, partridge berry, and green adder's mouth. The majority of the uplands are pine–mixed hardwoods. Several upland swamp forests, dominated by swamp-gum, are present.

On the eastern side of the park are salt marshes with smooth cordgrass, needle rush, and saltmarsh aster bordered by salt shrub thickets with marsh-elder, sea ox-eye, seaside goldenrod, and yaupon.

The roadsides harbor many species of the ruderal gardens.

CLARENDON COUNTY

Bennett's Bay Heritage Preserve: Site 20

Bennett's Bay is protected by the Nature Conservancy of South Carolina and access is controlled. One must contact the Columbia office (803-254-9049) for directions and permission to enter.

Bennett's Bay is a Carolina bay. A pocosin community occurs in the bay proper and the sand ridges on the southeastern side harbor the xeric sandhills community. The rare lamb-kill grows in the ecotone between the pocosin and xeric community. Typical pocosin species occur in the bay including red bay, loblolly bay, sweet bay, honeycup, and fetterbush. The xeric sandhills community is dominated by turkey oak and scattered longleaf pine.

Bennett's Bay was formally called Junkyard Bay and appears as the latter on topographic and county road maps. It is located 2 miles northwest of Foreston between U.S. Highway 521 and County Road 211. A sometimes hard-to-find dirt road leads off 521 to the bay; in wet weather it should not be used by vehicles.

COLLETON COUNTY

Ashepoo River: Site 21

The Ashepoo River passes through the heart of the ACE Basin. One may enter the river at a boat landing on U.S. Highway 17 and follow it all the way to Otter Island (see St. Helena Sound Heritage Preserve below) in St. Helena Sound. Along the way the river passes through tidal freshwater and brackish marshes, and finally into the salt marshes of the sound. The freshwater portion of the river passes through abandoned rice fields of old plantations that support tidal freshwater marshes. The following freshwater marsh plants can be seen: indigo-bush, spider-lily, arrow arum, cat-tail, swamp rose, pickerelweed, large marsh-pink, swamp rose mallow, water willow, cardinal flower, and eryngo.

Colleton State Park: Site 22

Colleton State Park lies along the Edisto River at the town of Canadys off U.S. Highway 15. The park offers a short wildflower walk along a nature trail flanking the blackwater Edisto River and an old logging canal. The nature trail runs in the ecotone between the bottomland, bald cypress swamp of the river floodplain and the adjacent uplands. Trees of the swamp community are bald cypress, tupelo gum, red maple, river birch, water ash, and water-elm (*Planera aquatica*); trees of the uplands include white oak (*Quercus alba*), dogwood, and horse sugar. Shrubs of the swamp include Virginia willow, swamp dogwood, titi, and elderberry; fetterbush and sparkleberry are upland shrubs seen along the trail. Swamp herbs include false nettle, aquatic milkweed, and water willow; upland herbs include heart-leaf and partridge berry. Look for the following vines along the trail: cross vine, cow-itch, climbing hydrangea, and the low, trailing *Smilax pumila*.

Crosby Oxypolis *Heritage Preserve: Site 23*

Crosby *Oxypolis* Heritage Preserve is a site for Canby's Dropwort (*Oxypolis canbyi*), a Federally Endangered Species. It also harbors a variety of natural gardens for wildflower trips. The Preserve lies in the triangle formed by the intersection of State Highway 63 and Secondary Road 191, northwest of Walterboro.

The dominant community of the Preserve is an upland swamp-gum swamp forest. The site originally contained pond cypress; however, the cypress was timbered years ago as evidenced by sawed-off stumps. Only a few, scattered, small specimens of cypress remain today. The swamp is easy to wade through when flooded (or it may be dry); the canopy is closed and shrubs are confined to old stumps and root masses. Shrubs include sweet bay, myrtle dahoon, fetterbush, red bay, and sweet pepperbush. Loblolly bay is present along with Spanish moss, coral greenbrier, and mistletoe.

The best way to reach to swamp forest is to park off Highway 63 at its intersection with Secondary Road 33. Look for the sign Pine Grove #1 Baptist Church. To the opposite side of Highway 63, follow the same line as Highway 33. Before you reach the swamp, you pass through a stand of swamp-gum, slash pine, and pond cypress. The shrub layer is dense, mostly because of the fetterbush. Species to look for in this zone are sweet bay, pipewort, yellow trumpet pitcher-plant, myrtle dahoon, coral greenbrier, and Virginia chain-fern (*Woodwardia virginica*).

To the northwest of the swamp and bordering Secondary Road 191 is a pocosin. Along 191 the pocosin is dense, but inward there are many openings where the shrubs are thin. This is where Canby's dropwort occurs. Absence of fire is endangering the population of dropwort. Species to look for in the pocosin and ecotone are red chokeberry, sphagnum, fetterbush, sweet bay, pipewort, and bamboo-vine.

Combahee River: Site 24

The Combahee is a coastal, blackwater river that was the center of the rice-growing industry in what is now part of the ACE Basin Focus Area. It provides ready access to all five islands of the St. Helena Sound Heritage Preserve (see below): Otter, Ashe, Beet, Big, and Warren islands. It forms the boundary between Beaufort and Colleton counties and passes by Laurel Springs and Longbrow plantations on the Colleton side and Clay Hall Plantation on the Beaufort side.

The most convenient landing is at Highway 17. One can either go upriver or downriver from this point. Downriver one initially passes through abandoned rice fields, some with intact banks and managed for waterfowl, and others with dikes broken and in various stages of returning back to swamp forests. Tidal freshwater marshes are extensive with the following species: cardinal flower, swamp rose, pickerelweed, native wisteria, tag alder, southern rein orchid (*Habenaria flava*), lizard's tail, water willow, indigo-bush, alligator-weed, spider-lily, arrow arum, marsh bulrush, eryngo, and climbing hempweed.

Regenerating tidal freshwater swamps occur in scattered tracts along both sides of the river. Trees and shrubs typical of this natural garden found on the Combahee include red maple, red bay, bald cypress, swamp-gum, water ash, river birch, Virginia willow, and sweet-gum.

Downriver the tidal freshwater marshes and swamps grade into brackish marshes with saw grass, southern wild rice, and arrow-leaf morning-glory. The brackish marshes grade into the coastal salt marshes of St. Helena Sound where the Combahee and Coosaw rivers merge.

Edisto Beach State Park: Site 25

Edisto Beach offers typical coastal dune and beach gardens. Severe erosion has eliminated most of the dunes; however, a sand ridge extends north of the island to Jeremy Inlet where one can view species of the dunes and beaches. The maritime forest has mostly been developed for park facilities.

The nature trail passes through a modified maritime forest, ending at the Indian midden at the edge of the salt marsh. The midden has eroded greatly because of the nearby creek. The only true calciphyte on this midden is shell-mound buckthorn; the rest of the plants are typical maritime species. One rare herb occurs along the trail in close proximity to the midden: incised groovebur (*Agrimonia incisa*). Only four populations are known in South Carolina. Numerous wildflowers occur along the trail including elephant's foot, coral bean, wild olive, cross vine, bull bay, and chinquapin.

Edisto Nature Trail: Site 26

Edisto Nature Trail in Jacksonboro is owned and maintained by the Timberlands Division of Westvaco Corporation. It is open to the public year around; no admission is charged. Well-marked trails are maintained with identification tags on trees and shrubs. Boardwalks cross wet areas. Free brochures are available that describe the trails and associated plants.

Most of the forests along the nature trails are secondary vegetation on abandoned agricultural fields, old rice fields, and phosphate mines. A wide variety of flowering trees and shrubs can be seen: yellow jessamine, black cherry, beauty-berry, red maple, tulip tree, flowering dogwood, devil's walking stick, southern crab-apple, yaupon, sycamore, red buckeye, hop hornbeam, and dahoon.

One can venture off the trails into tidal freshwater swamps and freshwater marshes along the Edisto River where bald cypress, swamp rose, pickerelweed, water hemlock, arrow arum, native wisteria, and groundnut grow.

Otter Island (St. Helena Sound Heritage Preserve): Site 27

St. Helena Preserve consists of five islands: Big, Warren, Ashe, Beet, and Otter. Of the five, Otter Island and its associated hummocks afford the greatest diversity of natural gardens. Otter is a barrier island and can easily be reached from three boat landings: (1) Edisto Marina at the southern end of Edisto Island (this is a commercial landing and requires a ramp fee); (2) Live Oak Landing on Big Bay Creek (take State Secondary Highway 1461 off State 174); or (3) the public landing at the end of State Secondary Road 26 (Bennetts Point Road) on the Ashepoo River. Or, as the end of a longer boat trip, one may come down either the Ashepoo or Combahee rivers into St. Helena Sound to Otter Island.

Otter Island and its associated hummocks are bounded by the ocean on the south, Fish and Jefford creeks on the east and north, and the Ashepoo River and St. Helena Sound on the west. Landing on the front beach may be difficult because of the ocean waves. The front beach can best be reached by going down the Ashepoo River and landing on the western side where it is sheltered. The Ashepoo River can be reached from Edisto Marina and Live Oak landing by travelling up the Edisto River to Fenwick Cut and then into the Ashepoo.

The dunes on Otter Island harbor a rich mix of plants: devil-joint, seaside pennywort, silver-leaf croton, sea oats, camphorweed, dune sandbur, dunes evening-primrose, seaside panicum, and trailing bluet. Maritime shrub thickets lace the dunes. The front beech is eroding, so the beech community is absent. The most well-developed maritime forests are on the eastern end of the island.

Landward to the barrier island are numerous smaller islands and hummocks. These can best be reached by Fish and Jefford creeks, or by two smaller creeks on the western side of Otter Island. *Spartina* marshes, salt flats, salt shrub thickets, and shell deposits are present. Sweet grass is present on the edge of the hummocks. Look for shell-mound buckthorn along the edge of hummocks on shell deposits.

The hummocks and smaller islands offer exceptional opportunities to view maritime forest species because the underbrush is thin and ingress is easy. Good stands of wild olive occur on several hummocks. Also look for saw palmetto which reaches its northern range on these hummocks. The pines are slash pine (*P. elliottii*).

Westvaco's Bluff Trail: Site 28

Westvaco's Bluff Trail is designed for use by educational groups, environmental groups, and others interested in modern forestry practices in the Lowcountry. The upland area is a managed forest and will change over time as modern forest management is applied by Westvaco. Other areas, such as the former inland rice fields now harboring a swamp community, will remain untouched.

The trail is located about 10 miles south of Walterboro at the junction of U.S. Highway 17A and Secondary Road 66. All visits are by appointment through Westvaco's South Area Office located in Walterboro. Call 803-538-8353. All visitors to the Bluff Trail will be accompanied by a forester or trained resource professional.

The heart of Westvaco's Bluff Trail is the inland rice field, once part of the vast estate of the Heyward family which had a rice-growing dynasty in the Combahee River area. Walking trails, in part using the old banks, give access to the rice fields. Here is a secondary, but mature, alluvial bald cypress–tupelo gum swamp forest. Other species that can be seen are red maple, supplejack, cross-vine, blue flag iris, butterweed, leucothoe, Virginia willow, bur-reed, mistletoe, resurrection fern, coral greenbrier, lizard's tail, arrow arum, and southern wild rice. In the spring the water is covered by floating bladderwort.

The entire Bluff Trail is 3 miles long; the rest of the trail passes through a mixture of secondary forests and pine plantations. Numerous species of the natural and ruderal gardens are present. Look for horse sugar, dogwood, yellow jessamine, Cherokee rose, dwarf azalea, Easter lily, elephant's-foot, heart-leaf, beauty-berry, winged sumac, mistflower, violet wood sorrel, Virginia creeper, pale gentian, false-dandelion, coral bean, broad-leaved eupatorium, giant beard grass, marsh bulrush, black-root, narrow-leaved skullcap, St. Peter's-wort, bull nettle, and Japanese honeysuckle.

DILLON COUNTY

Little Pee Dee State Park and Bay: Site 29

This park is the treasure of the Pee Dee area for wildflower enthusiasts. Six distinct natural gardens occur: xeric sandhills, pocosins, upland swamp forests, freshwater aquatics, freshwater marshes, and pine–mixed hardwoods.

The park lies along the eastern side of the Little Pee Dee River. Adjacent to the river, wind-blown sand deposited during the early formation of the coastal plain created a series of ridges and swales. The ridges support typical xeric sandhills species while the swales support pocosins and upland swamp forests. These communities lie along State Highway 22. Wildflowers in the pocosins include highbush blueberry, possum-haw, fetterbush, honeycup, sweet bay, and titi. Red maple, dog-hobble, and Virginia willow occur in the swamps.

Within the park, Beaver Pond Nature Trail ends at Beaver Pond, a walk of about a mile. Several freshwater aquatics occur in the open water of the pond: cow-lily, fragrant water-lily, and species of Lemnaceae. The edge of the pond harbors a freshwater marsh with arrow arum, common cat-tail, water willow, marsh bulrush, lizard's tail, and bur-reed.

The trail begins off the main road and passes through a loblolly pine plantation. Toward the latter half of the trail it passes through pine–mixed hardwoods.

DORCHESTER COUNTY

Francis Beidler Forest in Four Holes Swamp: Site 30

Beidler Forest is a National Audubon Society Sanctuary near Harleyville, South Carolina. It can be entered only from the southwest via Secondary Road 28 off U.S. Highway 178. From the east take I-26 west to Exit 187; go south (left) on S.C. 27 to U.S. 78; go west (right) on 78 to U.S. 178; follow the Beidler Forest signs from there to the sanctuary. From I-95 or the west, take I-26 east to Exit 177; go south (right) on S.C. 453 to U.S. 178; go east (left) on 178 through Harleyville and follow the signs to the sanctuary.

The heart of the 3,600-acre sanctuary is the largest remaining stand of original growth bald cypress–tupelo gum forest in the world. A 1½ mile-long boardwalk takes you through a portion of the alluvial swamp. Canoe trips can be arranged in season by reservation with the Sanctuary director (803-462-2150). At the Visitors Center one can view a slide program and other pictures and displays which will aid toward understanding and appreciating the swamp. Using this book, one can readily identify many of the wildflowers in Beidler Forest.

A rich variety of herbs can be seen from the boardwalk or from a canoe: golden-club, Spanish moss, green-fly orchid, resurrection fern, cardinal flower, false nettle, butterweed, obedient plant, aquatic milkweed, and lizard's tail. The stump-floating log microhabitat is common throughout the swamp with false nettle, skullcap, *Lycopus rubellus,* and *Hypericum virginicum* visible from the boardwalk.

Numerous flowering shrubs, vines, and trees are present along the boardwalk: Virginia willow, poison ivy, swamp dogwood, snowbell, red maple, leucothoe, cross vine, supple-jack, button-bush, and coral greenbrier.

Four Holes is a flowing swamp stream fed by springs and rain run-off from surrounding, higher areas. In the heart of Four Holes Swamp are ancient groves of bald cypress and tupelo gum, towering over clear pools and blackwater sloughs. Many of the cypress trees are believed to be six hundred years old.

One rare herb to note is Carolina trillium (*Trillium pusillum*), a candidate for endangered species status, which blooms in late March–April. The first part of the boardwalk passes through a pine–mixed hardwood community that was severely affected by Hurricane Hugo. Almost the entire canopy was downed. The site harbors the only known population of Carolina trillium in the state. In the past four years since the hurricane, the trillium has bloomed, apparently not being affected by the canopy loss.

The boardwalk also passes through another natural garden, the hardwood bottom. It is recognized by the presence of dwarf palmetto (*Sabal minor*).

Givhans Ferry State Park: Site 31

Givhans Ferry State Park lies adjacent to the Edisto River, a free-flowing blackwater river which begins above Aiken and winds through the southwestern section of the state before discharging into the Atlantic Ocean at Edisto Island. The park was the site of an old ferry which was the chief crossing of the Edisto River that connected Charleston to more inland areas of the state.

The park offers species of the marl and beech forests. Adjacent to the parking area in front of the Office Community Building is a deep ravine that cuts through a limestone bluff. Although the vegetation is dense, one can still reach the ravine and follow it to the river. Species that can be seen are hop hornbeam, widespread maiden fern (*Thelypteris kunthii*), thimbleweed, alumroot, pawpaw, redbud, sparkleberry, heart-leaf, dogwood, cross vine, bull bay, American beech, lopseed, white oak (*Quercus alba*), box elder (*Acer negundo*), and Christmas fern (*Polystichum acrostichoides*).

Up-river from the Community Building, the bluff forms a nearly vertical wall down to the river. Growing on the wall is a population of Venus'-hair fern (*Adiantum capillus-veneris*), an uncommon fern in the coastal area. The vertical wall, unfortunately, can only be seen from the river.

Along the river's edge is a narrow, alluvial floodplain supporting a swamp forest with a few species of note: river birch, river oats, water-elm (*Planera aquatica*), and bald cypress.

McAlhany Preserve: Site 32

McAlhany Preserve is owned by the Charleston Natural History Society—the Charleston Chapter of the National Audubon Society. It is open to the public only by permission of the Natural History Society which can be reached by phone listed under Audubon Society, Charleston Chapter (803-577-7100). The Preserve is located along State Highway 19 southwest of St. George. More precise directions can be obtained from the Audubon Chapter when visitation is requested. A visitor may also get directions to the trails and natural features of the Preserve.

McAlhany Preserve borders the Edisto River, a blackwater river. Along the river's edge are found bald cypress–tupelo gum swamp forests and hardwood bottoms. The rare sarvis holly (*Ilex amelanchier*) and green-fly orchid occur in the swamp forests. An ox-bow lake harbors freshwater aquatics.

The most significant site on the Preserve is a deciduous forest along a high bluff and adjacent, former floodplain. Both oak-hickory and beech forests are present. The beech forests are more common on the floodplain and the oak-hickory forests are more common on the bluff. The herbaceous flora is especially rich with the following species: bloodroot, pennywort, may-apple, little sweet betsy, heart-leaf, green-and-gold, cancer-root, Indian cucumber-root, downy rattlesnake plantain (*Goodyera pubescens*), spotted wintergreen, partridge berry, beech-drops, Indian pipe, Jack-in-the-pulpit, and Easter lily.

The deciduous forest begins to the right of the entrance gate, with the bluff beginning further into the woods.

GEORGETOWN COUNTY

Black River Swamp Heritage Preserve: Site 33

The Black River Swamp Preserve is located along the Black River, a beautiful blackwater river that begins near Bishopville in Lee County and ends a 100 miles distant at Winyah Bay near

Georgetown. The Preserve is approximately 30 miles from the coast, near Andrews; the river is tidal at this point. To reach the Preserve from Andrews, take S.C. Highway 41 north to Secondary Road 38; turn right on 38 and go for 1.8 miles where you will see the sign Pine Tree Landing. Turn right onto the dirt road which leads to Pine Tree public boat landing. This is a modern, well-kept landing capable of handling large boat trailers. The Preserve is on the southern side of the river and extends for 3 miles; the boundary is clearly marked with Heritage Preserve signs.

The Preserve consists of 1,276 acres and harbors tidal freshwater swamps, tidal freshwater marshes, and loblolly pine hummocks. Two creeks, Lester and Puncheon, branch from the river and provide better access to the adjacent swamps. Although the river is tidal, the current is gentle and canoeists should have no trouble putting in and taking out at the same ramp.

Trees and shrubs to look for in the swamps that border the river and creeks include bald cypress, tupelo gum, red maple, sweet-gum, water ash, and river birch. Mistletoe is common, and all three epiphytes of the Lowcountry are found: green-fly orchid, Spanish moss, and resurrection fern. Shrubs and vines to look for include native wisteria, tag alder, Virginia willow, cross vine, climbing hydrangea, leucothoe, supplejack, coral greenbrier, poison ivy, and dwarf palmetto.

These tidal swamps were too high up the river to have been converted to rice fields; therefore, the tidal marshes are not extensive and are confined to narrow strips along the river's edge and at the base of trees. Look for obedient plant, climbing aster, southern wild rice, water parsnip, and climbing hempweed.

Brookgreen Gardens Xeric Sandhills: Site 34

Brookgreen Gardens harbors an outstanding example of the xeric sandhills garden. Visitors must make an appointment with the staff to visit the site by calling 803-237-4218.

This is one of the few sites in the Lowcountry and Pee Dee to have rosemary (plate 281). Other xeric species include longleaf pine, turkey oak, tread-softly, Carolina ipecac, thistle, sandhills milkweed, reindeer moss, dwarf huckleberry, and gerardia.

Huntington Beach State Park: Site 35

The causeway to the beach passes through two marshes: a salt marsh and a freshwater lagoon made by damming the upper reach of the salt marsh. A boardwalk takes visitors into the salt marsh. The lagoon harbors numerous freshwater aquatics and freshwater marsh species: duckweeds, mosquito fern (*Azolla caroliniana*), common cat-tail, and sacred bean; arrow-leaf morning-glory occurs along the edge.

From the picnic area at the northern end of the island, a

trail leads to the front beach. The dunes harbor the typical coastal dune species. In the freshwater swales behind the dunes grow seaside goldenrod and common marsh-pink. Maritime shrub thickets occur behind the dunes. The maritime forests are not well-developed and are too dense to penetrate easily.

Samworth Wildlife Management Area and Great Pee Dee River: Site 36

Access to the Samworth Wildlife Management Area is from State Highway 52. Here a boat ramp gives access to the Great Pee Dee River and various creeks, some of which connect to the Waccamaw River. Vast areas of tidal freshwater marshes occur along and between these rivers and creeks. Once mostly tidal freshwater swamps, the swamps were converted to rice fields during the 1800s. Today tidal freshwater marshes dominate the abandoned fields; the river banks harbor patches of tidal freshwater swamp forests. Prominent marsh and swamp species include swamp rose, pickerelweed, elderberry, swamp dogwood, southern wild rice, cow-lily, tag alder, indigo-bush, leather-flower, and Virginia willow.

U.S. Highway 701 between Plantersville and Yauhannah: Site 37

This section of U.S. Highway 701 passes through longleaf pine flatwoods, pocosins, and longleaf pine savannas. Numerous species of these communities occur along the roadside during the spring, summer, and fall. The adjacent land is private, and there are few house sites or commercial developments along this section of 701.

Frog's britches can be seen in flower during the spring along the ditch banks. Summer species include pine lily, pineland hibiscus, Barbara's-buttons, false asphodel, smooth meadow-beauty, orange milkwort, pipewort, and giant white-topped sedge.

HAMPTON COUNTY

Webb Wildlife Center: Site 38

The Webb Wildlife Center is one of the premier wildflower sites available to the public in the Lowcountry; over six hundred species of vascular plants have been identified on the Webb Center. The Center comprises 5,866 acres that border the Savannah River and extends 6 miles inland. It is open to the public year around; however, during the deer season (October–January 1) and the turkey season (April) public access is controlled. Visitors should call the S.C. Department of Natural Resources at Garnett, 803-625-3569, to make arrangements to use the Webb Center. Overnight accommodations and boat trips can be arranged. A brochure covering activities available at the Webb Center can be obtained from the DNR.

To reach the Webb Center, turn west off of U.S. Highway 321 in Garnett onto State 20 and go for approximately 2 miles until you come to the Webb Center sign on the left.

Select timbering and prescribed burning of the pine forests have created excellent wildflower conditions. The many roads of the Webb Center traverse extensive stands of both longleaf pine flatwoods and savannas and loblolly pine forests; oak-hickory forests and pine–mixed hardwoods interlace the pinelands. Two natural lakes, Bluff and Flat, remnants of former river-beds, support bottomland bald cypress–tupelo gum swamp forests on their margins. The floating-log community occurs in both lakes. Numerous freshwater aquatics occur in the lakes. Two artificial lakes also support a variety of freshwater aquatics including sacred bean. A 2-mile trail leads from Bluff Lake to the Savannah River; it passes through mature hardwood bottoms laced with wetland sloughs supporting bottomland swamp forests.

Special plants to look for include ladies'-eardrops along the margin of Bluff Lake, spreading pogonia in the longleaf pine savannas, and hairy wicky and spiked medusa (a rare orchid) in the grassy, longleaf pine flatwoods.

HORRY COUNTY

Cartwheel Bay Heritage Preserve: Site 39

Cartwheel Bay Heritage Preserve is difficult to locate, and persons visiting the site for the first time should obtain a map from the Heritage Trust Program. The following directions are taken from their guide to Cartwheel Bay. From Conway, take U.S. Highway 701 to the fork of 701 and State Road 410. Go left on 401 to Playcards Crossroads and County Road 19, turn left on 19 and continue for about 8.3 miles. You should now see a large, redwood-colored building on the right side of the road. This is the Cartwheel Bay Community Center. Proceed beyond this point for about .2 mile, and you will see a red building on the left, a small access road after the building, and another red house next. The access road is the road on the left between the two buildings.

Cartwheel Bay is a 680-acre Carolina bay complex, with one large 150-acre bay and five smaller ones. Natural gardens include pocosins, longleaf pine flatwoods, longleaf pine savannas (both toothache and wiregrass), oak savannas, and xeric sandhills. Fire has maintained the open nature of the area for years.

The longleaf pine–toothache grass savannas are especially spectacular with numerous orchids and carnivorous plants including snowy orchid, crested fringed-orchid, white fringed-orchid, yellow fringed-orchid, hooded pitcher-plant, sweet pitcher-plant, frog's britches, and yellow trumpet pitcher-plant. Many other flowering herbs of the longleaf pine savannas occur year around. In the ecotones between the pocosins and longleaf pine savannas occur two rare species—pixie moss and Venus' fly trap.

Lewis Ocean Bay Heritage Preserve: Site 40

Lewis Ocean Bay Heritage Preserve consists of approximately 9,300 acres. To reach the Preserve, take S.C. Highway 90 off U.S. Highway 501 east of Conway. Travel approximately 6 miles north to International Road (formally Burroughs Road) directly across from Wild Horse Subdivision. Turn east onto International Road, take a left at the first fork which is Kingston Road, and follow Kingston Road until it dead ends into Target Road. The entrance to Lewis Ocean Bay is located to the right.

Lewis Ocean Bay is so large that it is impractical to cover it in one trip. Numerous visits should be made at different growing seasons to fully appreciate its plant diversity. The following natural communities can be seen: pocosins, xeric sandhills, longleaf pine savannas (both types, one dominated by wiregrass and one dominated by toothache grass), and longleaf pine flatwoods. Many of the savannas and flatwoods were timbered and planted in loblolly pine before the Preserve was established.

Along the edge of the many Carolina bays of the Preserve are xeric sandhills that harbor the rare Carolina ipecac and typical species such as sandhills milkweed, tread-softly, reindeer moss, and sandhills baptisia. The pocosins harbor the rare leather-leaf and frog's britches and numerous ericaceous species such as the spectacular honeycup. Between the pocosins and sand ridges are ecotones that harbor creeping blueberry, Venus' fly trap, and sundews. The pine savannas harbor native orchids such as bearded grass-pink and spreading pogonia and carnivorous plants such as hooded pitcher-plant, butterworts, and sundews.

Little Pee Dee River Heritage Preserve (Vaughn Tract): Site 41

The Vaughn Tract can be reached from State Highway 917 which runs between Mullins in Marion County and Loris in Horry County. The tract lies along the eastern side of the Little Pee Dee River. A dirt road leads off Highway 917 through an iron gate on the northeast side of the highway and runs parallel to the river through the entire length of the 3,771-acre tract.

The heart of the Preserve is the series of fluvial sand ridges interspaced between pocosins or bay forests. The ridges support the xeric sandhills community. Species of note include reindeer moss, prickly-pear, longleaf pine, turkey oak, stylisma (*Bonamia pickeringii*), wire plant (*Stipulicida setacea*), joint weed (*Polygonella polygama*), wiregrass, gerardia, and an unusual beak rush, *Rhynchospora megalocarpa*.

The ecotones between the sand ridges and pocosins support creeping blueberry and the rare pixie moss and lamb-kill. Good populations of these rare species can be found in the ecotones along the first sand ridges after turning off Highway 917.

Extensive pocosins occur throughout the tract. Look for loblolly bay, sweet bay, red bay, fetterbush, bamboo-vine, and titi.

Myrtle Beach State Park: Site 42

The front dunes of Myrtle Beach State Park were highly eroded during Hurricane Hugo. Restoration efforts, using a fence and discarded Christmas trees, succeeded in creating stable dunes by the summer of 1991. Typical beach and dune plants became established and provide a good representation of this garden. Russian thistle and sea rocket occur on the beach and sea oats, beach pea, gaillardia, dune spurge, sand ground-cherry, silver-leaf croton, camphorweed, and dune sandbur can be found on the dunes.

The maritime forests immediately behind the dunes suffered considerable damage from the winds of Hurricane Hugo, and pines and other trees were killed by the salt water. Inland, on the southern end, mature maritime forests, a mixture of maritime and oak-hickory species, provide a sparse mix of wildflowers: yaupon, bull bay, beauty-berry, and devil's walking stick.

Waccamaw River Preserve: Site 43

Waccamaw River Preserve was obtained in 1987 by the Nature Conservancy of South Carolina. It is open to visitors without reservations and is located at the confluence of the Waccamaw River and Intracoastal Waterway. Two boat landings allow access to the site via either the Waccamaw River or Intracoastal Waterway: Enterprise Landing at the end of State Highway 926 or Peachtree Landing at the end of State Highway 611.

The 689-acre tract consists primarily of tidal freshwater swamps and tidal freshwater marshes. The tidal swamps are the type that were converted to rice fields; there is no evidence, however, that this tract was ever cultivated. The lack of large bald cypress indicates that the swamps were timbered years ago. Marsh wildflowers include water-spider orchid, arrow arum, pickerelweed, spider-lily, obedient plant, ground-nut, eryngo, and cardinal flower. The tidal swamp forests harbor bald cypress, red maple, tupelo gum, water ash, green-fly orchid, Virginia willow, aquatic milkweed, swamp rose, and snowbell.

JASPER COUNTY

New River: Site 44

Public access to the New River is a boat ramp on State Highway 170. The New River is one of the most desolate and scenic rivers along the Carolina coast. Virtually no signs of inhabitation occur along its entire course to the coast (10 miles). The upper reaches of the river support tidal freshwater swamp forests and tidal freshwater marshes. Downriver the freshwater system gives way to brackish marshes which then grade into the salt marshes near the ocean. North of the landing is a continuation of the tidal freshwater swamps and marshes; the river narrows here, however, and ingress is limited to smaller boats.

Abandoned rice fields and canals are scattered along the upper, freshwater section of the river. Secondary tidal freshwater swamps have reclaimed the fields. Swamp plants include swamp-gum (*Nyssa biflora*), red maple, sweet-gum, button-bush, leather-flower, Virginia willow, poison ivy, water ash, laurel oak (*Quercus laurifolia*), wax myrtle, and dwarf palmetto (*Sabal minor*).

The tidal freshwater marshes are rich in species and include pickerelweed, eryngo, native wisteria, swamp rose, seashore mallow, lizard's tail, cardinal flower, climbing hempweed, and large marsh-pink.

The brackish marshes are distinct and include southern wild rice, wild rice, saw-grass, and arrow-leaf morning-glory.

S.C. Highway 119: Site 45

The 6-mile section of State Secondary Road 119 from where it joins State Secondary Road 201 to its intersection with State Primary Road 119 contains some of the rare and common xeric species found in the Tillman Sand Ridge Heritage Preserve (see below) plus numerous species of the sandy, dry, open woods. Some rare xeric species to look for include gopher apple, warea, soft-haired coneflower, and rose dicerandra. Other xeric species include gopherweed, thistle, sandhills milkweed, and sandhills baptisia. Some open woods species include blue star, wild pink, cottonweed, and blazing-star.

Tillman Sand Ridge Heritage Preserve: Site 46

Tillman Preserve is located 5 miles west of Tillman on State Secondary Road 119. The 953-acre Preserve harbors three natural communities: the alluvial bald cypress–tupelo gum swamp forests and hardwood bottoms along the brownwater Savannah River, and the xeric sandhills community. The sandhills community is the core of the Preserve and consists of fluvial sand ridges formed along the Savannah River during prehistoric times. Longleaf pine and turkey oak form the tree canopy; xeric herbs include wiregrass (*Aristida stricta*), gopherweed, gopher apple, sandhills milkweed, sandhills baptisia, warea, soft-haired coneflower, summer-farewell, woolly golden-aster, cottonweed, wild pink, blue star, and rose dicerandra.

MARION COUNTY

Little Pee Dee River Heritage Preserves: Site 47

The Little Pee Dee River is the county line for Marion and adjacent Horry County from its junction with the Lumber River (just below Nichols in Marion County) to its confluence with the Great Pee Dee River. The Heritage Trust Program has acquired four tracts of land south of State Highway 917, to approximately 8 miles south of U.S. Highway 501. Three tracts—the Dargan, Ward, and Tilghman—lie in Marion County, while the Huggins tract lies in Horry County. All four tracts front the Little Pee Dee River for approximately 15 miles.

These preserves are best seen via boat; put in at Highway 917 and travel downstream, or put in at U.S. 501 and go upstream to the Ward and Tilghman tracts or downstream to the Dargan and Huggins tracts. Visitors must understand that private property is interspaced between the heritage property. Either watch for signs denoting the heritage property or carry an adequate map to the properties.

Three natural communities grace this pristine and winding, blackwater river: stream banks, alluvial bald cypress–tupelo gum swamp forests, and alluvial hardwood bottoms.

Behind the bars and along the exposed edge occur swamp forests with a wide variety of species such as cardinal flower, cow-itch, bald cypress, tupelo gum, Virginia willow, red maple, snowbell, and purple lobelia. The rare sarvis holly (*Ilex amelanchier*), with its red drupes, can be seen during the fall as a high, much-branched shrub in the swamp forests.

ORANGEBURG COUNTY

Santee State Park: Site 48

Santee State Park consists of 2,496 acres and is located on the edge of Lake Marion in the town of Santee. The park is open year around and has excellent sites for wildflowers. Most of the 2,496 acres is covered in upland, pine–mixed hardwoods; however, one site harbors the deciduous forest gardens which are a must for wildflower observation.

The deciduous forests occur around the limestone caverns. Because of the presence of a rare bat (the southeastern myotis [*Myotis austroriparius*]) in the caverns, the area is restricted, and visitors must get permission from park officials (and directions) to visit the site.

Three deciduous gardens are present: marl, oak-hickory, and beech forests. The key to locating the three gardens is the limestone cave which park officials will show visitors. To locate the oak-hickory forests, follow the eastern fork of the stream that comes out of the limestone cave and forms Chapel Branch. Chapel Branch, which is a small lake, runs east, and the mature, oak-hickory forests occur on the sloping bluff adjacent to the branch. Species of note are white oak (*Quercus alba*), southern red oak (*Q. falcata*), mockernut hickory, redbud, horse sugar, coral honeysuckle, flowering dogwood, cancer-root, sparkleberry, red buckeye, sassafras, flowering spurge, green-and-gold, yellow jessamine, black cherry, fringe-tree, elephant's-foot, Indian pink, crab-apple, lyre-leaved sage, and needle grass (*Stipa avenacea*).

The western fork of the stream leading from the cave eventually leads into a ravine with an exposed limestone bluff on the southern side. Elements of the marl and beech forests merge, making it difficult to separate the two communities. Typical marl forests species are southern sugar maple (*Acer saccharum* subsp. *floridanum*), meadow parsnip, and alumroot. Liverleaf (*Hepatica americana*), a rare coastal species, occurs on the bluff. Species more characteristic of the beech forests are American beech, witch-hazel, bloodroot, may-apple, little sweet betsy, heart-leaf, green-and-gold, sweet shrub, tulip tree, spotted wintergreen, windflower (*Thalictrum thalictroides*), strawberry bush (*Euonymus americanus*), spotted wintergreen, and crane-fly orchid (*Tipularia discolor*). Along the stream edge are species such as Jack-in-the-pulpit, pawpaw, climbing hydrangea, Easter lily, golden-club, and spicebush (*Lindera benzoin*).

Glossary

The plant structures marked with an asterisk are illustrated in figure 5.

Accretion. Geologically, a slow addition to land by deposition of sediment by natural forces such as wind or water.

ACE Basin. Lower part of the watershed of the Ashepoo, Combahee, and Edisto rivers.

Achene. A small, dry, one-loculate, one-seeded, indehiscent fruit; for example, in the Asteraceae family.

Adventitious. Developing in an unusual or irregular position, such as roots from stems.

Aerenchyma. Aerating tissue in aquatic plants, characterized by large intercellular spaces; functions in flotation of the plant and storage and diffusion of gasses.

Alien. From another country; exotic.

Alluvial soil. Soil developing from recent alluvium (material deposited by running water); exhibits no horizon development; typical of floodplains.

*Alternate leaves.** Only one leaf at a node.

Annual. Plant growing from seed to fruit in one year, then dying.

*Anther.** The pollen producing part of the stamen.

Aphrodisiac. Stimulator of sexual desire.

Appressed. Lying flat against.

Aromatic. Having a fragrant, sweet-smelling, or spicy aroma.

Ascending. Growing obliquely upward at about a 40–60 degree angle from the horizontal.

Asexual reproduction. Reproduction that does not involve fusion of gametes; for example, fragmentation of a rhizome with each fragment growing into a new plant.

Asphyxia. Unconsciousness or death resulting from lack of oxygen.

Astringent. Agent that causes contraction of tissues, thereby lessening secretion.

Auricle. Any earlike lobe or appendage.

Axil. The angle formed by the upper side of a leaf and the stem from which it grows.

Axillary. In an axil.

Barrier island. Offshore body of land that rises permanently above normal sea level; for examples, Otter Island and Capers Island.

Basal leaves. Leaves at the base of the stem.

Berm. Backshore terrace lying between beach and dunes. See plate 1.

Berry. A simple, fleshy, usually indehiscent fruit with one or more seeds; for example, a tomato or grape.

Biennial. Living for two years, then dying naturally.

*Bipinnately compound leaf.** Twice pinnate, the primary leaflets once-again pinnate.

*Blade.** Flattened and expanded part of a leaf.

Bog. In general terms, a wide variety of soggy, moist, spongy areas; in the strictest usage, reserved for only raised peatlands.

Bract. Modified leaf, usually smaller than a foliage leaf, often situated at the base of a flower or inflorescence.

Bulb. An underground, fleshy enlargement of stem and leaves, as in the onion.

Buttress. Additional, often flattened, supporting tissue at the base of the trunk, as in bald cypress.

Calcareous. Made up of, having, or typical of calcium carbonate, calcium, or limestone; for example, calcareous soil.

Calciphyte. A plant living in soil abundantly supplied with calcium ions.

*Calyx.** The collective term used to describe all the sepals of a flower.

Canopy. The top layer of leaf growth within most woody communities.

Capsule. A dry dehiscent fruit with more than one chamber; it may open by pores, by splitting vertically, or by the top coming off like a lid.

*Catkin.** A spikelike inflorescence bearing either male or female flowers, as in willows and oaks.

Chlorophyll. The green pigment of plants.

Ciliate. Beset with a marginal fringe of hairs.

Clasping. A leaf whose base wholly or partly wraps around or surrounds the stem.

Colonial. Adjective of colony.

Colony. Growing in clumps produced asexually from underground structures such as rhizomes, rootstocks, stolons, or roots.

Column. In orchids, a structure formed by the union of stamens, style, and stigma; the supporting structure of the hood in pitcher-plants.

Community. Group of interacting plants and animals inhabiting a given area.

Composite. Any member of the Asteraceae family.

*Compound leaf.** Leaf in which the blade is subdivided into two or more leaflets.

Conifer. Any of the cone-bearing gymnosperms, such as pines.

Consumption. Old term for tuberculosis.

Cone-bearing plants. Technically, the gymnosperms; plants that produce seeds not enclosed by an ovary, as in pine trees.

Corm. A thickened, vertical, underground stem with thin, scalelike leaves.

*Corolla.** All the petals of a flower, separate or united; the inner whorl of the perianth.

Cross-pollination. Pollination between two different plants of the same species.

Cultivated plant. A purposely grown plant; it may be a native plant moved from local woodlands into a garden or yard, or a plant from another country.

Cuticle. Waxy, noncellular layer on outer surface of epidermal wall of plant organs (mostly leaves) that prevents excessive water loss.

Deciduous. Falling away, not persistent or evergreen.

Decoction. An extraction of a plant made by boiling a plant part in water.

Decomposition. Breakdown of complex dead organic substances into simpler ones necessary to recycle minerals.

Dehiscent. Opening by pores or slits to discharge the contents.

Dermatitis. Inflammation of the skin.

Detritus. Accumulated mass of partially decomposed remains of animals and plants which forms in aquatic systems.

Dichotomous. Two-forked, the branches equal or nearly so.

Disjunct. A population of plants growing far from its main range.

*Dissected leaf.** The blade cut into, more or less, fine divisions.

Disseminate. To scatter widely, as in sowing seed.

*Divided leaf.** Any blade cut into divisions reaching three-fourths or more of the distance from the margin to the midvein or to the base.

Doctrine of Signatures. Primitive belief that the key to people's use of plants was hidden in the form of the plant itself, as in the red juice of bloodroot to treat blood disorders.

Dredged soil disposal site. Site used to dispose of soil dredged to maintain harbor or river depths; in the coastal area most often a section of banked marsh.

Drupe. A stone fruit; fleshy fruit with the single seed covered by a hard covering (stone).

Ecotone. Transitional area between two different communities, having characteristics of both, yet having a unique character of its own.

*Elliptic.** Widest in the center and narrowed to two equal ends.

Emergent. Aquatic plant with its lower part submerged and its upper part extended above water.

Emergent, nonpersistent. Emergent marsh species that fall to the surface of the water at the end of the growing season.

Emergent, persistent. Emergent marsh species that usually remain standing at least until the beginning of the next growing season.

Emersed. Rising above the surface of the water; applies to leaves.

Emetic. An agent that induces vomiting.

Endemic. Restricted to a small area or region.

Endosperm. The nutritive tissue of most seeds.

*Entire leaf.** Leaf margin without teeth, lobes, or divisions.

Ephemeral. Lasting only a short time.

Epiphyte. A plant growing on another plant but obtaining no nutrition from it; often referred to as an air plant, as in Spanish moss.

Essential oil. Volatile oils with characteristic odor, composed of various constituents and contained in plant organs.

Estuary. An area where fresh water and sea water meet and mix.

*Even-pinnately compound.** Said of compound leaves having an even number of leaflets; easily determined because there is a pair of leaflets terminally.

Evergreen. Bearing green leaves throughout the year; holding live leaves over one or more winters until new ones appear.

Exotic. Not native; from another country.

Expectorant. Substance that promotes expulsion of mucous from the respiratory tract.

Fascicle. A small bundle or tuft, as of leaves.

Febrifuge. Agent which relieves or reduces fever.

Fibrous roots. Root system composed of a mass of fiberlike roots with no main root predominating.

*Filament.** Part of the stamen that bears the anther.

Filiform. Threadlike; long and very slender.

Fire, crown. A fire that runs through the tops of trees.

Fire, ground. Fire in which organic matter in the soil is consumed, especially in peaty soils; fire that may burn for long periods entirely below the surface of the ground.

Fire, surface. A fire that burns only surface litter and small vegetation.

Flatwoods. Poorly drained, low-lying, nearly level timberland.

Fleshy. A plant having tissue that serves to store moisture, such as a cactus.

Flora. Collective term to refer to all plants of a area; a book dealing with the plants of an area.

Flowering plants. Technically the Angiosperms, plants that produce seeds enclosed in an ovary.

Fluvial. Caused by the action of flowing water.

Forb. Herbaceous plant other than a grass, sedge, or rush.

Frond. In Lemnaceae, the expanded leaflike stem which functions as a leaf.

Fruit. The seed-bearing structure of the plant; a matured ovary with its contents, often with attached parts.

Gland. A secreting surface or structure, or an appendage having the general appearance of such an organ.

Glandular. Bearing glands.

Habitat. Place where a plant or animal lives.

Halophyte. Plant able to survive and complete its life cycle in high salinity.

Haustorium. A bridge of xylem (figure 3) between host and parasite in plants through which water, minerals, and limited amounts of food pass from host to parasite.

*Head.** A dense inflorescence of sessile or subsessile flowers on a short or broadened axis, as in the Asteraceae.

Hemiparasite. Dependent on the host for water and minerals; contains chlorophyll and can make its own food, but may receive some food from the host, as in chaff-seed (*Schwalbea americana*).

Herb. Having no persistent woody stem above ground; also, a plant used in seasoning.

Herbaceous. Having the characteristic of an herb.

Hip. The fleshy to leathery hollow fruit of roses.

Holoparasite. Parasite completely dependent on the host for water, minerals, and food, as in beech-drops (*Epifagus virginiana*).

Humus. Organic material derived from partial decay of plant and animal matter.

Hydric. Habitats characterized by an abundant water supply.

Hydrophyte. A plant that grows in water.

Indehiscent. Remaining persistently closed; not opening by definite pores or sutures.

Indigenous. Native to the area.

Inflorescence. A flower cluster on a plant, or, especially, the arrangement of flowers on a plant.

Infusion. An extraction of a plant made by soaking the plant part in water.

Introduced. Plant brought intentionally from another area; such a plant may escape and become naturalized; for example, *Crotalaria spectabilis* brought as a green manure from India.

*Irregular flower.** A flower with petals that are not uniform in shape but are usually grouped to form upper and lower "lips."

Knee. Vertical outgrowth from the lateral roots of trees growing on soil subjected to long periods of inundation; the function is unknown; for example, bald cypress.

*Lanceolate.** Lance-shaped, much longer than wide and broadest near the base.

*Leaflet.** One of the leaflike parts of a compound leaf.

Legume. A dry fruit from a single ovary usually dehiscent along two sutures.

Lenticles. Small openings in the bark of roots and stems of flowering plants used for gas exchange.

Lichen. Unique composite organism formed by a symbiotic relationship between some sac fungi (and, to a lesser extent, club fungi) and a photosynthetic partner, either a blue green or green alga.

Limestone. Sedimentary rock composed mostly of carbonate of calcium.

*Linear.** Narrow and elongate with essentially parallel sides.

*Lip.** The lower petal of some irregular flowers, often showy, as in the orchids.

Litter. Accumulated mass of partially decomposed remains of plants (and animals) that collects on the forest floor.

*Lobed leaf.** Blade divided into parts separated by rounded sinuses extending one-third to one-half the distance between the margin and the midrib.

Locule. Cavity of an ovary or an anther.

Maritime. Located on or close to the sea.

Marl. Sedimentary rock formation composed of unconsolidated mixture of 35–65 percent carbonate of calcium and 65–35 percent clay.

Marsh. Wetland dominated by emergent, herbaceous vegetation.

Marsh, brackish. Marsh flooded regularly or irregularly by water of low salt content.

Marsh, freshwater. Marsh saturated or flooded with fresh water.

Marsh, high saltwater. Salt marsh flooded irregularly by spring tides or storm surges.

Marsh, low saltwater. Salt marsh flooded daily by tides.

Mesic. Moist but well-drained soils.

Mesophyte. Plant adapted to a mesic environment.

Mucilage. A substance of varying composition produced in cell walls of plants; hard when dry, swelling, and slimy when moist.

Native plant. One that originated in the area where it grows.

Naturalized. Plant from another area thoroughly established in a new area by being able to successfully reproduce naturally; for example, white clover from Europe now naturalized throughout the southeastern United States.

Natural selection. Natural process which results in survival of the best adapted individuals of a species and elimination of individuals less well adapted to their environment.

Nectar. Sweet substance secreted by special glands (nectaries) in flowers and in certain leaves.

Nitrogen fixation. Conversion of atmospheric nitrogen to forms usable by most organisms.

Node. Point on a stem where one or more leaves are borne or attached.

Nut. Indehiscent, one-seeded fruit having a hard outer wall, as in oaks and hickories.

*Oblanceolate.** Lanceolate, and attached at the narrow end.

Obligate. Limited to one mode of life or action, as an obligate parasite.

Oblique. Sides unequal, especially the base of a leaf.

*Obovate.** Ovate, and attached at the narrow end.

*Odd-pinnately compound leaf.** Said of compound leaves having an odd number of leaflets, this usually is easily determined because there is a single, terminal leaflet.

*Opposite leaves.** Two leaves inserted at the node opposite each other on the stem.

Ornamental. A plant cultivated for its beauty.

Outcrop. A stratum or formation, as of limestone or marl, that protrudes above the soil.

*Ovary.** The basal, enlarged part of the pistil that contains the ovules or seeds.

*Ovate.** Egg-shaped, and attached at the broad end.

*Ovule.** The structure that develops into the seed.

*Palmately compound leaf.** The leaflets diverge from a common point at the end of the leaf stalk.

*Panicle.** A compound inflorescence in which the main axis is branched one or more times and may support spikes, racemes, or corymbs.

Parasitic. Deriving food or mineral nutrition, or both, from another living organism.

*Peltate.** Attached to the stalk inside the margin, as in species of *Hydrocotyle.*

Perennial. A plant lasting for three or more years.

Perfect flower. A flower having both female (pistil) and male (stamen) parts.

*Perfoliate.** Describes those stalkless leaves whose base surrounds the stem, the stem thus apparently passing through it.

*Perianth.** The calyx and corolla collectively; the calyx alone if the corolla is absent.

Persistent. Remaining attached; not falling off.

Perturbation. Another word for "disturbance" with the added meaning of altering the state or direction of change in a system.

*Petal.** One of the individual parts, separate or united, of the corolla.

Pharmacopoeia. Book containing an official list of drugs along with recommended procedures for their preparation and use.

Photosynthesis. Synthesis of carbohydrates from carbon dioxide and water by chlorophyll-containing plants using light as energy and releasing molecular oxygen as a by-product.

*Pinnately compound leaf.** With the leaflets of a compound leaf placed on either side of the rachis, featherlike.

*Pistil.** The central, seed-bearing organ of a flower, usually composed of stigma, style, and ovary; the female part of a flower.

Pith. The central portion of a dicot stem.

Polymorphic. With three or more forms, such as the entire, two-lobed, or three-lobed leaves in sassafras.

Pome. A simple, fleshy fruit like an apple in which the flesh is derived from nonfloral parts.

Prickle. A small, usually slender outgrowth of the young bark.

Propagule. Any of the various structures of plants capable of developing into a new individual.

Prostrate. Lying flat on the ground; if a stem, may or may not root at the nodes.

Pubescence. A general term for hairs or trichomes.

Punctate. With translucent or colored dots, depressions, or pits scattered over the surface.

Pungent. Affecting the organs of smell or taste with a strong, acrid sensation.

Purge. To cause evacuation of the bowels.

Quinine. Main drug for treatment of malaria, derived from bark of species of *Cinchona* native to South America.

*Raceme.** A simple, indeterminate inflorescence of stalked flowers borne on a single, more or less elongated axis.

*Rachis.** The central elongated axis of an inflorescence or a compound leaf.

*Receptacle.** The base of the flower where all flower parts are attached.

Resinous. With the appearance of resin; glandular dotted.

Rhizome. Underground stem, usually horizontally orientated; sometimes functions in food storage.

Rhizomatous. Bearing rhizomes.

Rhizosphere. The soil immediately surrounding the root system.

Rootstock. An erect, rootlike stem or branch under or sometimes on the ground.

Rosette. Arrangement of leaves radiating from a crown or center, usually at or close to the ground.

Saline. Of, relating to, or containing salt; salty.

Samara. A dry, indehiscent, winged fruit, as in red maple (plate 139).

Saprophyte. Any organism that derives its nourishment from dead or decaying organic matter.

Savanna. A flat area with widely spaced trees, usually dominated by grasses.

Secondary forest. The forest occupying a site where the original forest was removed.

*Seed.** The matured ovule consisting of an embryo, seed coats, and stored food.

*Sepals.** One of the parts of the calyx, either separate or united.

Serotinous cones. Cones that remain on the tree several years and require the heat of fire to open them to release the seeds, as in pond pine (*Pinus serotina*).

Sessile. Without a stalk of any kind, as a sessile leaf.

Shade-tolerance. Capacity of a tree to develop and grow to maturity in the shade.

Sheath. A tubular envelope, usually used for that part of the leaf of a sedge or grass that envelopes the stem.

Shrub. A woody plant that remains low and produces several shoots or trunks from the base.

*Simple leaf.** A leaf with a blade in a single part, although it may be variously divided.

*Spadix.** A spike with a fleshy axis in which the flowers are embedded, as in Jack-in-the-pulpit.

*Spathe.** A large bract enclosing an inflorescence.

*Spike.** An elongated, indeterminate inflorescence of sessile or subsessile flowers.

Spine. A sharp-pointed, rigid, deep-seated outgrowth from the stem, not pulling off with the bark, as in Cactaceae.

Spring tide. Tide of maximum rise occurring near times of new and full moon.

Spur. A tubular or saclike projection from a petal or sepal.

*Stamen.** The pollen-producing organ of a flower; the male part of a flower.

*Standard.** Uppermost petal in a pea flower; also called the banner.

*Stigma.** The part of the pistil that receives the pollen; usually hairy or sticky.

Stolon. A slender stem that runs along the surface of the ground, or just below, and produces a new plant at the tip.

Stoma (plural *stomata*). Small opening on the surface of a leaf through which gaseous exchange takes place with the atmosphere.

*Style.** The elongated portion of the pistil that connects stigma and ovary.

Subcanopy. The layer of leaves just below the canopy in a forest.

Submerged. Growing entirely under water.

Sucker. Lateral underground shoot which leaves the roots or rhizomes and roots itself, forming an independent, individual plant.

Swamp. Forested, freshwater wetland often with saturated soil or standing water.

Symbiosis. The living together of two or more species, as in the fungal-algal symbiosis in the lichens.

Taproot. A large, elongated root, usually vertical.

Tendril. A slender twining or clasping structure that enables plants to climb.

Terrestrial. Living on the ground.

Trailing. Prostrate but not rooting.

Transpiration. Loss of water vapor from aerial parts of land plants.

Tree. A perennial woody plant of considerable stature at maturity and with one or few main trunks.

*Trifoliolate leaf.** A compound leaf with three leaflets.

Tuber. Fleshy, thickened, short, usually subterranean stem having numerous buds called "eyes," such as the potato.

Tuft. A dense clump, especially of bushes or trees.

Twining. Ascending by coiling around a support.

Two-ranked. When alternate leaves appear on just two sides of a stem.

*Umbel.** A flat-topped or rounded inflorescence having flowers on stalks of nearly equal length and attached to the summit of the peduncle, the characteristic order of blooming being from the outside toward the center.

Valve. One of the parts or segments into which a dehiscent fruit splits, as in a legume.

Vascular plant. Any of various plants of the Division Tracheophyta, typified by a conducting and supporting system of xylem and phloem; includes the ferns, cone-bearing, and flowering plants.

Velamen. A specialized moisture-absorbing tissue.

Vermifuge. An agent that expels worms from the intestine.

Vine. A plant which climbs by tendrils or other means, or which creeps or trails on the ground.

Water, brackish. Water that has salt concentration greater than fresh water and less than sea water.

Weed. In the broadest sense, a plant growing in a place it is not wanted. Usually weeds are aggressive colonizers of disturbed areas, and are frequently introduced; they are generally noxious, of no economic value, and compete with agricultural crops.

*Whorled leaves.** Three or more leaves inserted at one node.

*Wing.** A thin, flat extension found at the margins of plant parts; the lateral petal in a pea flower.

Xeric. Dry soils and sites, or adapted to dry conditions.

Xerophyte. Plant adapted to a xeric environment.

FIGURE 5: ILLUSTRATIONS OF PLANT STRUCTURES

This section presents shapes and arrangements of basic flower and leaf parts. The illustrations will help in understanding terminology presented in the text. Each item is defined in the glossary.

Leaf Shapes

Linear

Elliptic

Lanceolate

Olanceolate

Ovate

Obovate

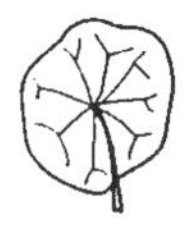

Peltate

Perfoliate

Leaf Parts

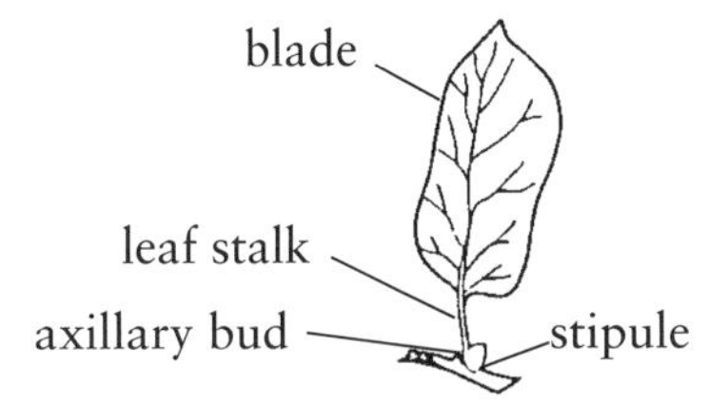

Parts of a Simple Leaf

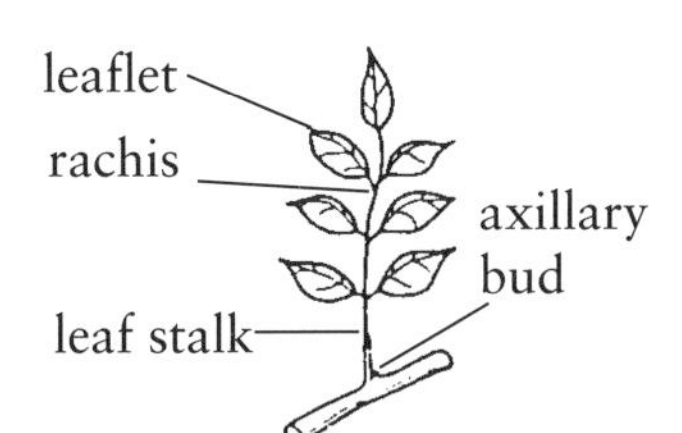

Parts of a Compound Leaf

Leaf Arrangement

Alternate

Opposite

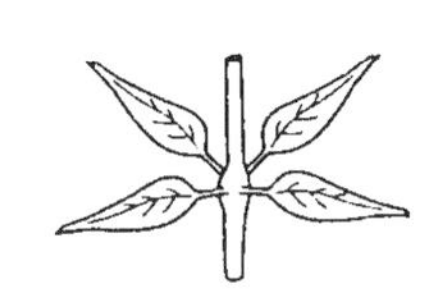

Whorled

Types of Compound Leaves

Palmately Compound

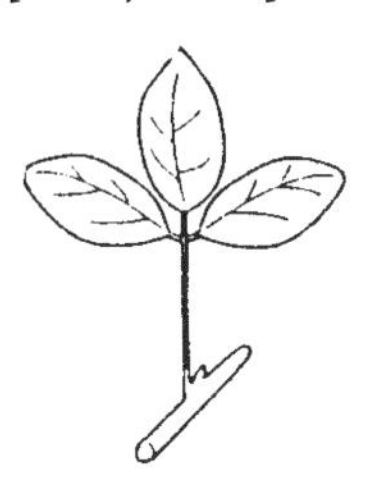

Trifoliolate

Odd-pinnate

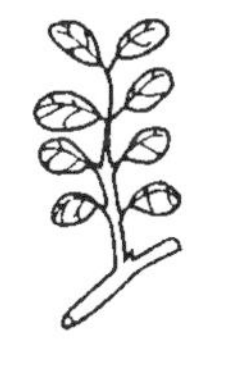

Even-Pinnate

Bipinnate

Leaf Margins

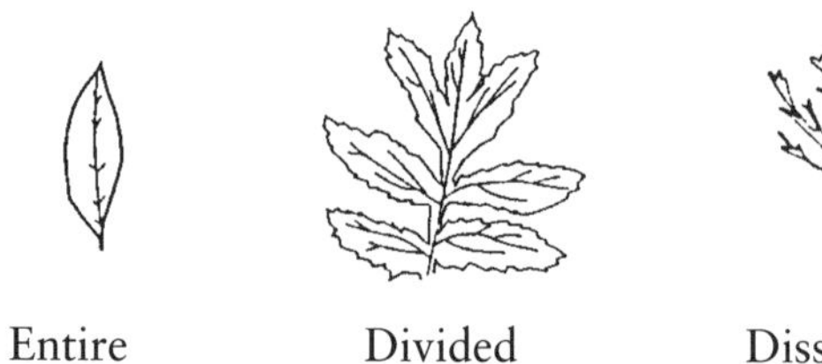

Parts of a Generalized Flower

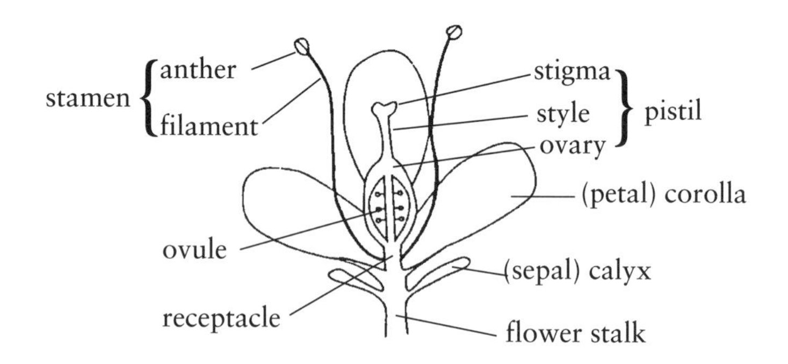

Irregular Flowers

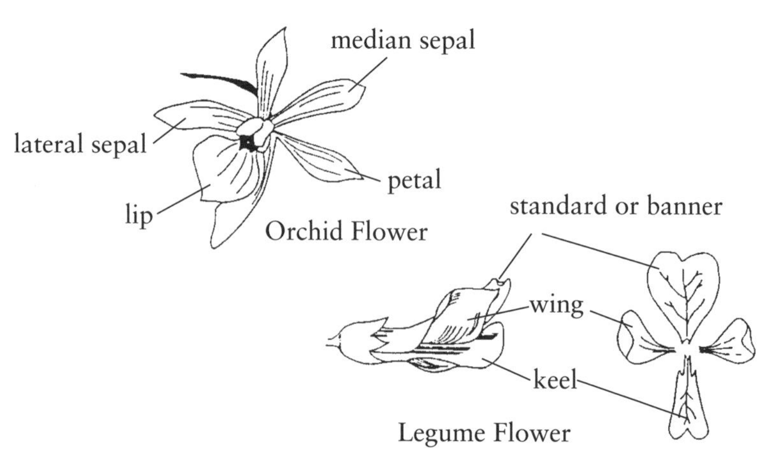

Types of Inflorescences

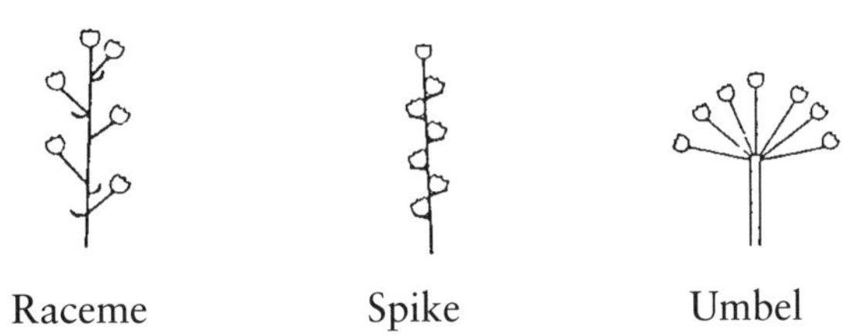

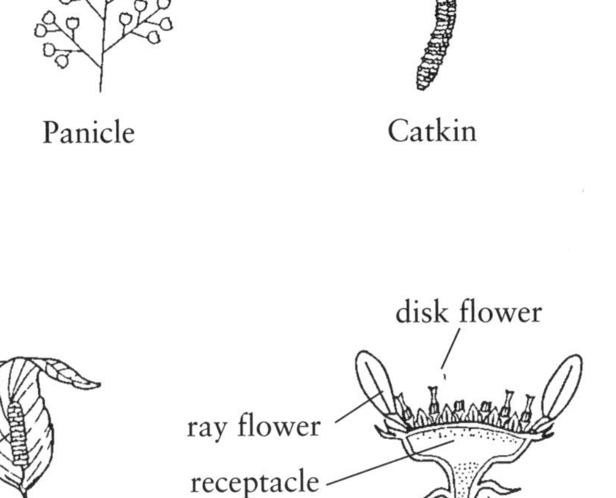

Bibliography

This bibliography lists literature cited in the text as well as other suggested references. Entries are grouped by topic (such as "Weeds") or, when more pertinent, by type (such as "Exhibit Books").

Aquatic Plants

Beal, Ernest O. 1977. *A Manual of Marsh and Aquatic Vascular Plants of North Carolina*. North Carolina Agricultural Experiment Station Technical Bulletin, No. 247.

Godfrey, Robert K., and Jean W. Wooten. 1979. *Aquatic and Wetland Plants of the Southeastern United States: Monocotyledons*. Athens: University of Georgia Press.

———. 1981. *Aquatic and Wetland Plants of the Southeastern United States: Dicotyledons*. Athens: University of Georgia Press.

Biographies of Botanists

Haygood, Tamara M. 1987. *Henry William Ravenel, 1814–1887*. Tuscaloosa: University of Alabama Press.

Rembert, David H. 1980. *Thomas Walter: Carolina Botanist*. Columbia: South Carolina Museum Commission Bulletin, no. 5.

Savage, Henry, Jr., and Elizabeth J. Savage. 1986. *André and Francois André Michaux*. Charlottesville: University Press of Virginia.

Carnivorous Plants

Folkerts, George W. 1982. "The Gulf Coast Pitcher Plant Bogs." *American Scientist* 70, no. 3: 260–70.

Lloyd, Francis E. 1976. *The Carnivorous Plants*. New York: Dover Publications.

Schnell, Donald C. 1976. *Carnivorous Plants of the United States and Canada*. Winston-Salem: John F. Blair.

Slack, Adrian. 1980. *Carnivorous Plants*. Cambridge: MIT Press.

Carolina Bays

Melton, Frank A., and William Schriever. 1933. "The Carolina Bays: Are They Meteorite Scars?" *Journal of Geology*

41: 52–66.
Johnson, Douglas. 1942. *The Origin of the Carolina Bays.* New York: Columbia University Press.
Porcher, Richard D. 1966. *A Floristic Study of the Vascular Plants in Nine Selected Carolina Bays in Berkeley County, South Carolina.* Master's thesis, University of South Carolina.
Savage, Henry, Jr. 1982. *The Mysterious Carolina Bays.* Columbia: University of South Carolina Press.

Cultivated and Ornamental Plants

Bailey, L. H. 1949. *Manual of Cultivated Plants.* 2d ed. New York: MacMillan Co.
Batson, Wade T. 1984. *Landscape Plants for the Southeast.* Columbia: University of South Carolina Press.
Briggs, Loutrel W. 1951. *Charleston Gardens.* Columbia: University of South Carolina Press.
Shaffer, Edward T. H. 1963. *Carolina Gardens.* 3d ed. New York: Devin-Adair Co.

Economic and Cultural

Hill, Albert F. 1952. *Economic Botany.* New York: McGraw-Hill Book Co.
Rosengarten, Dale. 1986. *Row Upon Row: Sea Grass Baskets of the South Carolina Lowcountry.* Columbia: McKissick Museum, University of South Carolina.
Simpson, Beryl B., and Molly C. Ogorzaly. 1986. *Economic Botany: Plants in Our World.* New York: McGraw-Hill Book Co.
Wood, Virginia S. 1981. *Live Oaking: Southern Timber for Tall Ships.* Boston: Northeastern University Press.

Edible Plants

Angier, Bradford. 1974. *Field Guide to Edible Wild Plants.* Harrisburg, Penn.: Stackpole Books.
Berglund, Berndt, and Clare E. Bolsby. 1971. *The Edible Wild.* New York: Charles Scribner's Sons.
Brown, Tom, Jr. 1985. *Tom Brown's Guide to Wild Edible and Medicinal Plants.* New York: Berkley Books.
Fernald, Merritt L., and Alfred C. Kinsey. 1958. *Edible Wild Plants of Eastern North America.* New York: Harper and Row.
Gibbons, Euell. 1962. *Stalking the Wild Asparagus.* New York: David McKay Co.
———. 1966. *Stalking the Healthful Herbs.* New York: David McKay Co.
Hall, Alan. 1976. *The Wild Food Trailguide.* New York: Holt, Rinehart and Winston.
Harris, Ben C. 1968. *Eat the Weeds.* Barre, Mass.: Barre Publishing.

Endangered, Threatened, and Rare Vascular Plants

Hardin, James W., and Committee. 1977. Vascular Plants. In J. E. Cooper, S. S. Robinson, and J. B. Funderburg, eds. *Endangered and Threatened Plants and Animals of North Carolina.* Raleigh: North Carolina State Museum of Natural History.

Rayner, Douglas A., and Committee. 1979. *Native Vascular Plants Endangered, Threatened, or Otherwise In Jeopardy in South Carolina.* Columbia: South Carolina Museum Commission Bulletin, No. 4.

Exhibit Books

Blagden, Tom, Jr. 1992. *South Carolina's Wetland Wilderness: The Ace Basin.* Englewood, Colo.: Westcliffe Publishers.

Blagden, Tom, Jr., Jane Lareau, and Richard D. Porcher. 1988. *Lowcountry: The Natural Landscape.* Greensboro, N.C.: Legacy Publications.

Ferns

Dunbar, Lin. 1989. *Ferns of the Coastal Plain.* Columbia: University of South Carolina Press.

Wherry, Edgar T. 1964. *The Southern Fern Guide.* New York: Doubleday & Co.

Folk Remedies and Medicinal Plants

Foster, Steven, and James A. Duke. 1990. *A Field Guide to Medicinal Plants.* Boston: Houghton Mifflin Co.

Hudson, Charles M., ed. 1979. *Black Drink: A Native American Tea.* Athens: University of Georgia Press.

Krochmal, Arnold, and Connie Krochmal. 1973. *A Guide to the Medicinal Plants of the United States.* New York: Quadrangle/New York Times Book Co.

Lewis, W. H., and M. P. F. Elvin-Lewis. 1977. *Medical Botany: Plants Affecting Man's Health.* New York: John Wiley and Sons.

Mitchell, Faith. 1978. *Hoodoo Medicine: Sea Islands Herbal Remedies.* Reed, Cannon and Johnson Co.

Morton, Julia F. 1974. *Folk Remedies of the Low Country.* Miami: E. A. Seemann Publishing.

Porcher, Francis P. 1869. *Resources of the Southern Fields and Forests.* Charleston: Walker, Evans and Cogswell.

Historical Books

Drayton, John M. 1802. *A View of South Carolina.* Charleston: W. P. Young (Reprint Co., Spartanburg, 1972).

Ramsey, David. 1858. *History of South Carolina.* Newberry: W. J. Duffie (Reprint Co., Spartanburg, 1960).

Manuals and Floras

Batson, Wade T. 1984. *Genera of Eastern Plants.* 3d ed., rev. Columbia: University of South Carolina Press.

Conquist, A. J. 1980. *Vascular Flora of the Southeastern United States*. Vol. 1: *Asteraceae*. Chapel Hill: University of North Carolina Press.

Radford, Albert E., Harry E. Ahles, and C. Ritchie Bell. 1968. *Manual of the Vascular Flora of the Carolinas*. Chapel Hill: University of North Carolina Press.

Rickett, Harold W., and The New York Botanical Garden. 1967. *Wild Flowers of the United States*. Vol. 2: *The Southeastern States*. New York: New York Botanical Garden and McGraw-Hill Book Co.

Small, John K. 1933. *Manual of the Southeastern Flora*. Chapel Hill: University of North Carolina Press.

Strausbaugh, P. D., and Earl L. Core. 1977. *Flora of West Virginia*. Grantsville: Seneca Books.

Native Orchids

Correll, Donovan S. 1978. *Native Orchids of North America*. Stanford: Stanford University Press.

Gupton, Oscar W., and Fred C. Swope. 1986. *Wild Orchids of the Middle Atlantic States*. Knoxville: University of Tennessee Press.

Luer, Carlyle A. 1972. *The Native Orchids of Florida*. Brooklyn: New York Botanical Garden.

———. 1975. *The Native Orchids of the United States and Canada Excluding Florida*. Brooklyn: New York Botanical Garden.

Natural History Guides

Barry, John M. 1980. *Natural Vegetation of South Carolina*. Columbia: University of South Carolina Press.

Braun, E. Lucy. 1950. *Deciduous Forests of Eastern North America*. Philadelphia: Blakistin Co.

Dennis, John V. 1988. *The Great Cypress Swamps*. Baton Rouge: Louisiana State University Press.

Duncan, Wilbur H., and Marion B. Duncan. 1987. *Seaside Plants of the Gulf and Atlantic Coasts*. Washington: Smithsonian Institution Press.

Lyons, Janet, and Sandra Jordan. 1989. *Walking the Wetlands*. New York: John Wiley & Sons.

Nelson, John B. 1986. *The Natural Communities of South Carolina*. Columbia: Technical Report, South Carolina Wildlife and Marine Resources Department.

Porcher, Richard D. 1985. *A Teacher's Field Guide to the Natural History of The Bluff Plantation Wildlife Sanctuary*. New Orleans: Kathleen O'Brien Foundation.

Rayner, Douglas A., et al. 1984. *Inventory of Botanical Natural Areas in Colleton, Dorchester, Horry and Jasper Counties, South Carolina*. Columbia: Technical Report, South Carolina Wildlife and Marine Resources Department.

Smith, Richard M. 1989. *Wild Plants in America*. New York: John Wiley & Sons.

Wells, B. W. 1932. *The Natural Gardens of North Carolina*. Chapel Hill: University of North Carolina Press.

Parasitic Flowering Plants

Kuijt, Job. 1969. *The Biology of Parasitic Flowering Plants.* Berkeley: University of California Press.

Plant Ecology

Daubenmire, R. F. 1959. *Plants and Environment.* New York: John Wiley and Sons.

Oosting, Henry J. 1956. *The Study of Plant Communities.* 2d ed. San Francisco: W. H. Freeman and Co.

Poisonous Plants

Kingsbury, John M. 1964. *Poisonous Plants of the United States and Canada.* Englewood Cliffs: Prentice-Hall.

Westbrooks, Randy G., and James W. Preacher. 1986. *Poisonous Plants of Eastern North America.* Columbia: University of South Carolina Press.

Popular Wildflower Books

Batson, Wade T. 1987. *Wild Flowers in the Carolinas.* Columbia: University of South Carolina Press.

Bell, C. Ritchie, and Bryan J. Taylor. 1982. *Florida Wild Flowers.* Chapel Hill: Laurel Hill Press.

Brown, Clair A. 1972. *Wildflowers of Louisiana and Adjoining States.* Baton Rouge: Louisiana State University Press.

Duncan, Wilbur H., and Leonard E. Foote. 1975. *Wildflowers of the Southeastern United States.* Athens: University of Georgia Press.

Hunter, Carl G. 1984. *Wildflowers of Arkansas.* Little Rock: The Ozark Society Foundation.

Justice, William S., and C. Ritchie Bell. 1968. *Wild Flowers of North Carolina.* Chapel Hill: University of North Carolina Press.

Martin, Laura C. 1989. *Southern Wildflowers.* Atlanta: Longstreet Press.

Niering, William A., and Nancy C. Olmstead. 1979. *The Audubon Society Field Guide to North American Wildflowers (Eastern Region).* New York: Alfred A. Knopf.

Timme, S. Lee. 1989. *Wildflowers of Mississippi.* Jackson: University Press of Mississippi.

Rice Culture

Allston, Robert F. W. 1846. "Rice." *Commercial Review of the South and West* 1: 320–57.

Doar, David. 1970. *Rice and Rice Planting in the South Carolina Low Country* (2nd printing). Charleston: Charleston Museum.

Hawley, Norman R. 1949. "The Old Rice Plantations in and around the Santee Experimental Forest." *Agricultural History* 23: 86–91.

Heyward, Duncan C. 1937. *Seed from Madagascar.* Chapel

Hill: University of North Carolina Press.

Hillard, Sam B. 1975. "The Tidewater Rice Plantation: An Ingenious Adaptation to Nature." *Geoscience and Man* 12: 57–66.

Porcher, Richard D. 1985. "Rice Culture." In Richard D. Porcher, *A Field Guide to The Bluff Plantation Wildlife Sanctuary*. New Orleans: Kathleen O'Brien Foundation.

———. 1987. "Rice Culture in South Carolina: A Brief History, The Role of the Huguenots, and Preservation of its Legacy." *Transactions of the Huguenot Society* 92: 1–22.

Salley, A. S., Jr. 1919. "The Introduction of Rice into South Carolina." *Bulletin of the Historical Commission of South Carolina* 6. Columbia: State Co.

Scientific Journal Articles

Gaddy, L. L. 1982. "The Floristics of Three South Carolina Pine Savannas." *Castanea* 47: 393–402.

Hunt, Kenneth W. 1943. "Floating Mats on a Southeastern Coastal Plain Reservoir." *Bulletin of the Torrey Botanical Club* 70, no. 5: 481–88.

Hunt, Kenneth W. 1947. "The Charleston Woody Flora." *American Midland Naturalist* 37: 670–756.

Porcher, Richard D. 1981. "The Vascular Flora of the Francis Beidler Forest in Four Holes Swamp, Berkeley and Dorchester Counties, South Carolina." *Castanea* 46: 248–60.

Trees, Shrubs, and Vines

Brown, Claud L., and L. Katherine Kirkman. 1990. *Trees of Georgia and Adjacent States*. Portland: Timber Press.

Coker, William C., and Henry R. Totten. 1945. *Trees of the Southeastern States*. Chapel Hill: University of North Carolina Press.

Duncan, Wilbur H. 1975. *Woody Vines of the Southeastern United States*. Athens: University of Georgia Press.

Duncan, Wilbur H., and Marion B. Duncan. 1988. *Trees of the Southeastern United States*. Athens: University of Georgia Press.

Elias, Thomas S. 1980. *The Complete Trees of North America*. New York: Times Mirror Magazines.

Peattie, Donald C. 1966. *A Natural History of Trees of Eastern and Central North America*. New York: Bonanza Books.

Walker, Laurence C. 1990. *Forests: A Naturalist's Guide to Trees & Forest Ecology*. New York: John Wiley & Sons.

———. 1991. *The Southern Forest: A Chronicle*. Austin: University of Texas Press.

Weeds

Martin, Alexander C. 1987. *Weeds*. New York: Golden Press.

Index of Common and Scientific Names

Numbers in italic type refer to color plates. Names in parentheses are not main entries, but are mentioned in the text. Synonyms are indicated by (Syn.). Scientific names are given in italics and common names in plain text.

General Index